Crown Gall

Advances in Understanding Interkingdom Gene Transfer

Edited by
Walt Ream and Stanton B. Gelvin

APS PRESS
The American Phytopathological Soc
St. Paul, Minnesota

This publication is based on presentations from the Sixteenth Annual Crown Gall Conference held on September 3 and 4, 1995, at the Friday Harbor, Washington, Laboratory. To make the information available in a timely and economical fashion, this book has been reproduced directly from computer-generated copy submitted in final form to APS Press by the editors of this volume. No editing or proofreading has been done by the Press.

Reference in this publication to a trademark, proprietary product, or company name by personnel of the U.S. Department of Agriculture or anyone else is intended for explicit description only and does not imply approval or recommendation to the exclusion of others that may be suitable.

Library of Congress Catalog Card Number: 96-86520
International Standard Book Number: 0-89054-222-8

Printed in the United States of America on acid-free paper

The American Phytopathological Society
3340 Pilot Knob Road
St. Paul, Minnesota 55121-2097, USA

Preface

On September 3-4, 1995, the Northwest Crown Gall Conference was held at Friday Harbor Laboratory to honor Eugene W. Nester and Milton P. Gordon for their pioneering discoveries in crown gall research and their outstanding service to the scientific community.

Speakers at this meeting were invited to contribute chapters to this volume, which documents recent work spanning the breadth of the crown gall field.

Acknowledgments

We thank the following sponsors, and the contact people responsible for securing these gifts, for their generous financial support.

Pioneer Hi-Bred International Inc. (Dr. Elizabeth Hood)

Monsanto (Dr. Robb Fraley)

Calgene (Dr. Vic Knauf)

Agracetus Inc. (Dr. Kenneth A. Barton)

USDA (Dr. R. James Cook and Dr. Steve Farrand)

Thanks also to Dr. Steve Winans for the idea to hold a meeting at Friday Harbor in honor of Gene Nester and Milt Gordon. Special thanks to Dr. Stan Gelvin for his help with planning, and thanks to Stan and Steve for their help with address lists.

We are indebted to Joyce Smith, Noel Bruzzichesi, Ray Bigler, and Jane Halsey, for their assistance with conference logistics.

The editors thank all of the authors, whose efforts made this volume possible. We greatly appreciate the work of Teri Warren and Lynette Bergeron, who undertook the monumental word processing task required to produce an attractive camera-ready book.

Many thanks to all who journeyed to share this exciting weekend.

Foreword

Less than two decades ago a research team led by Drs. Eugene W. Nester, Milton P. Gordon, and Mary-Dell Chilton published a watershed paper proving that *Agrobacterium tumefaciens* caused crown gall disease through stable transfer of specific bacterial genes to plant cells. This remarkable discovery catapulted interest in crown gall research into the stratosphere. It also assisted in the birth of a new field, plant molecular biology, which depends heavily upon the ability to create transgenic plants. Transgenic plant technology in turn led to the founding of the plant biotechnology industry, and today the general public enjoys the fruits of the first genetically engineered crops. Rarely do scientific discoveries have such a significant impact.

Drs. Nester and Gordon have remained at the fore of their field to this day, but their contributions are not restricted to advancement of scientific knowledge. They have trained a considerable number of scientists who have enjoyed remarkable success due not only to the training received from Drs.Nester and Gordon, but also due to the support they provided afterward. We are grateful for all you have done. Thank you, Gene and Milt.

Left to right: Mary-Dell Chilton, Milton Gordon, Eugene Nester

Table of Contents

CHAPTER 1

Signal Detection by VirA

Sharon Lafferty Doty*, Joe Don Heath, and Eugene W. Nester

Department of Microbiology, University of Washington, Seattle, WA 98195

*Present address: Department of Biochemistry, University of Washington, Seattle, WA 98195

ABSTRACT: VirA is the critical sensor protein which allows *Agrobacterium* cells to detect the presence of susceptible plant cells. In this paper, we present data showing major differences in the ability of the VirA protein to induce the *vir* genes in two different strains. Based upon data from the characterization of a variety of *virA* mutants, we present a model of how VirA may detect inducing sugars and phenolic compounds.

Introduction

VirA and VirG comprise a two-component regulatory system responsible for the activation of virulence (*vir)* gene expression in response to signals from a plant wound site (14). Two component systems consist of a sensor histidine kinase which regulates its activity in response to environmental signals; and a response regulator whose activity is controlled by the kinase. The kinase catalyzes transfer of a ϒ-phosphoryl group from MgATP to a conserved histidine residue, and the response regulator catalyzes the transfer of the phosphoryl group from the phosphorylated histidine in the kinase to its own conserved aspartate residue in its active site (35). Some kinases are known to be active as homodimers, and in some cases the phosphorylation of the histidine residues occurs in *trans* between monomers within a dimer (24,36,44,45).

VirA is a critical environmental sensor required for virulence. The short amino-terminal tail and most of the carboxy-terminus is cytoplasmic, while most of the amino-terminus is periplasmic, flanked by two transmembrane (TM) domains (21). Following the second transmembrane domain (TM2) are a kinase domain and receiver domain (7). Adjacent to TM2 is an amphipathic helix, an alpha helix with strong hydrophilic and hydrophobic faces (29). This structure may help anchor the protein in the membrane (30). The kinase domain is critical for tumorigenesis; mutation of the phosphorylatable histidine residue (His474) results in an avirulent phenotype (17). The receiver domain has sequence similarity to VirG over a limited region which includes the phosphorylatable Asp residue. This domain may provide an additional level of control, as in the ArcB/ArcA system (16), by requiring two signals, one causing phosphorylation of the receiver domain within VirA, and a second signal to phosphorylate VirG.

The signals which activate VirA include phenolic compounds such as acetosyringone (AS), monosaccharides such as glucose, and an acidic pH. The sugars signal VirA indirectly through the periplasmic sugar binding protein, ChvE (5). There may be a phenolic binding protein which interacts with VirA (18); however, recent genetic evidence suggests that the phenolic compounds interact directly with VirA (19). The VirA protein, an integral inner membrane protein, undergoes autophosphorylation and then transfers the phosphate to the VirG protein. VirG binds to a *vir* box in the promoters of all the *vir* genes and activates transcription.

Over 40 different phenolic compounds have been shown to act as *vir* gene inducers, although some inducing compounds are specific to certain strains. First to be isolated were acetosyringone and hydroxyacetosyringone from wounded *Nicotiana tabacum* (34). The structural requirements for the A6 VirA-activating phenolic compounds are a benzene ring with a hydroxyl group at the R4 position and a methoxyl constituent at the R-3 position (22). There is no requirement for the methoxyl group for phenolic compounds which can induce the *vir* genes of strain KU12 (19).

Certain monosaccharides act synergistically with AS to induce the *vir* genes (3,5,32). Low levels (5 μM) of AS do not normally induce the *vir* genes unless an inducing sugar is also present. There are thirteen known inducing monosaccharides, most of which are known components of plant cell walls. The structural features which these 13 sugars share are a six-membered pyranose ring, an equatorial hydroxyl at C-3 (with the exception of idose), and either equatorial or axial hydroxyls at C-2 and C-4 (axial is favored at C-2). *chvE* is a chromosomal virulence gene required for *vir* gene induction in response to sugars (5,15). Shimoda *et al.* reported that a *virA* mutant strain which no longer showed enhancement of *vir* gene induction in response to glucose was suppressible by a mutation in *chvE*, demonstrating genetically that VirA and ChvE interact directly (31).

Another factor involved in the induction of the *vir* genes is pH. A pH less than 6.0, like that of plant wound sap, is optimal (34), but the exact pH optimum varies between strains (39). Other factors include a temperature of approximately 20°C (1), a low level of inorganic phosphate (43), and the presence of opines (42).

Although the *virA* genes of different *Agrobacterium* strains are basically the same in terms of structure and function, significant differences exist. For example, the VirA protein of the limited host range (LHR) strains from grapevines is ineffective at directing *vir* gene induction in response to AS (20) except under specific conditions of low temperature and reduced inoculum size (39). Furthermore, some strains can transfer T-DNA into maize. Nopaline-type strains are able to transfer a modified T-DNA containing a viral genome into monocots, whereas octopine-type strains cannot. The determining factor was found to be the *virA* gene (28).

The goal of this research was to obtain a better understanding of how VirA detects and responds to plant signal molecules. We accomplished this by isolating and characterizing a number of VirA mutants with altered *vir* gene induction properties.

Materials and Methods

Bacterial strains. A348 is C58 with pTiA6 (12). MX1 is A348/*chvE::*Tn5 (15). At11054 is C58/*chvE::*Tn*5* (8). C58/*virA*- has a deletion of the first 2113 bp of the *virA* gene (27). At12004 is a C58/*chvE*$^-$*virA*$^-$ double mutant (11).

Media and chemicals. All *Agrobacterium* cultures were grown at 28°C in MG/L broth (9) or AB/Glycerol (10). Induction broth (IB) contained 62.5 mM phosphate buffer, AB salts (6), and 0.5% glycerol. The pH of the IB was 5.8 for all strains with C58 *virA* and pH 5.5 for all strains with A6 *virA*. Antibiotics were used at the following concentrations (in µg/ml): kanamycin, 100; carbenicillin, 100; gentamicin, 50. Acetosyringone (Aldrich Chemical Company) was prepared as a stock solution of 100 mg/ml in dimethylformamide.

Plasmids. pSM243cd has a *virB::lacZ* fusion in pVK102 (34). pTC116 has the *gbpR* and *chvE* genes from C58 in the pUCD2 derivative, pTC110 (8). pJD102W has the C58 *virA* gene in the pUCD2 derivative, pDMD2 (11). pSL50 has the A6 *virA* gene in pUCD2 (11). pJD113 contains a hybrid *virA* gene with the periplasmic domain of C58 *virA* in A6 *virA*. pJD105W contains the reverse hybrid: the PPD of A6 *virA* in C58 *virA* (13).

β-galactosidase assays for *vir* gene induction. Strains were grown overnight in 2 ml AB/Glycerol, pelleted, and resuspended in IB, and 100 µl were used to inoculate 2 ml of the different IB solutions. After 16 hrs, β-galactosidase activity was assayed as described (11,33).

Results

C58 VirA requires ChvE for activation. Since we were interested in the how VirA detects inducing sugars through the periplasmic sugar binding protein ChvE, we chose to use the wild-type strain, C58. Previous studies of the VirA/ChvE interaction had been done using hybrid strains with the chromosome of C58 and the Ti plasmid of various octopine strains. Since VirA is pTi-encoded and ChvE is chromosomally-encoded, the hybrid interaction is not native. We began our study by comparing the effects of disruption of the *chvE* gene in C58 and in the hybrid strain A348. As shown in Figure 1 (11), the A348/*chvE* mutant, MX1, is still capable of *vir* gene induction in response to high levels of acetosyringone (AS); however, the synergistic response to low levels of AS with glucose is absent. The situation is quite different in the C58/*chvE* mutant, however. In At11054 where ChvE is not present, *vir* genes are not induced by

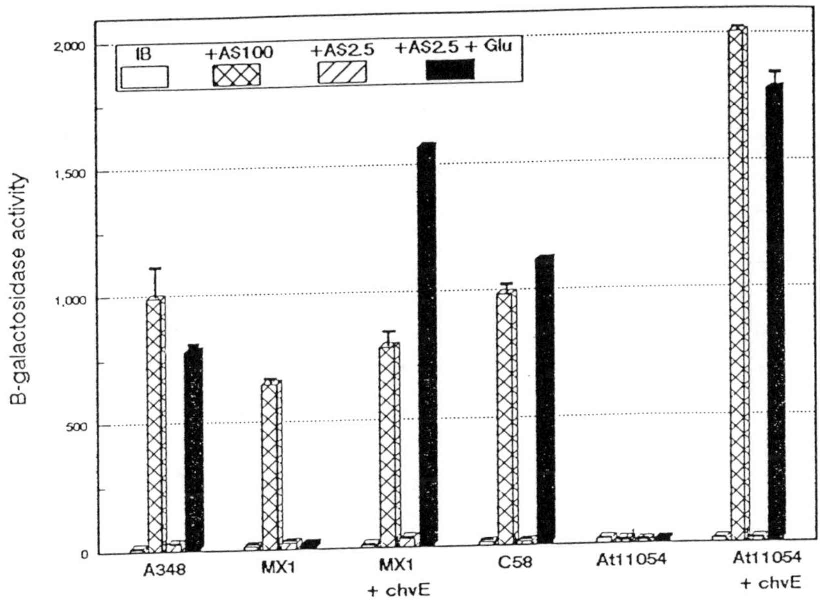

Figure 1. Expression of *virB::lacZ* in *chvE* mutants. Strains A348 (pTC110), MX1(pTC110), MX1(pTC116), C58(pTC110), At11054(pTC110), and At11054(pTC116), all containing pSM243cd, were incubated in IB with either 100 μM AS, 2.5 μM AS, or 2.5 μM AS with 10 mM glucose. Reprinted with permission from Doty, *et al.* 1996 (11).

even high levels of AS. Induction is restored in the mutant when *chvE* is provided *in trans*. These data demonstrate that 1) C58 VirA requires ChvE to induce the *vir* genes, and suggests that 2) ChvE, with or without its sugar ligand, interacts with VirA, since ChvE functions in *vir* gene induction even in the absence of glucose.

C58/A6 hybrid VirA molecules are activated by AS^{100} in the absence of ChvE. In an attempt to determine which domain of C58

VirA is responsible for the lack of induction, two hybrid *virA* genes (13) were introduced into the *chvE virA* double mutant, At12004. As shown in Figure 2, when the C58 VirA periplasmic domain (PPD) is replaced with that of A6, the *vir* genes are induced in response to AS^{100} in the absence of ChvE, suggesting that the C58 VirA PPD is responsible for the lack of induction. However, when the PPD of A6 VirA is replaced by that of C58, AS^{100} also induces. Therefore, there is a combination of effects from both the periplasmic and non-periplasmic domains.

Mutational analysis of C58 VirA. As we recently reported (11), we isolated 14 C58 *virA* mutants with altered *vir* gene induction properties. Six mutants had single missense mutations in the periplasmic domain ranging from amino acids 88 to 210, which resulted in a lack of induction like the C58 *chvE* mutant. When the corresponding mutations were made in the A6 *virA* sequence, the resulting mutants had a specific defect in the *vir* gene response to glucose like the A348 *chvE* mutant. These data suggest that the periplasmic domain must be intact in order to respond to inducing sugars. Figure 3 (11) shows a model of how the VirA/ChvE

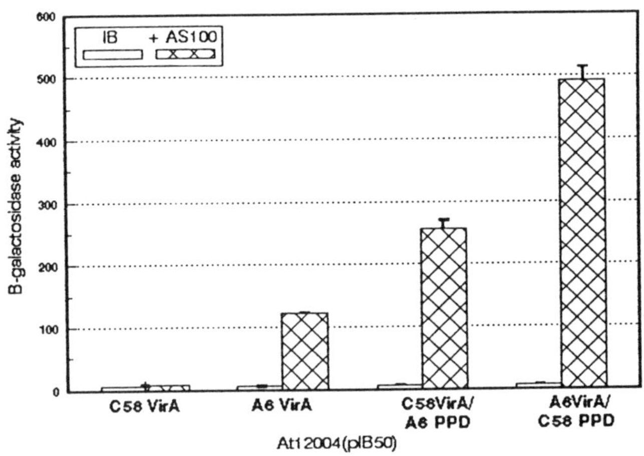

Figure 2. Expression of *virB::lacZ* in the C58/*virA chvE* double mutant At12004 (pIB50) containing the indicated *virA* genes on a separate plasmid.

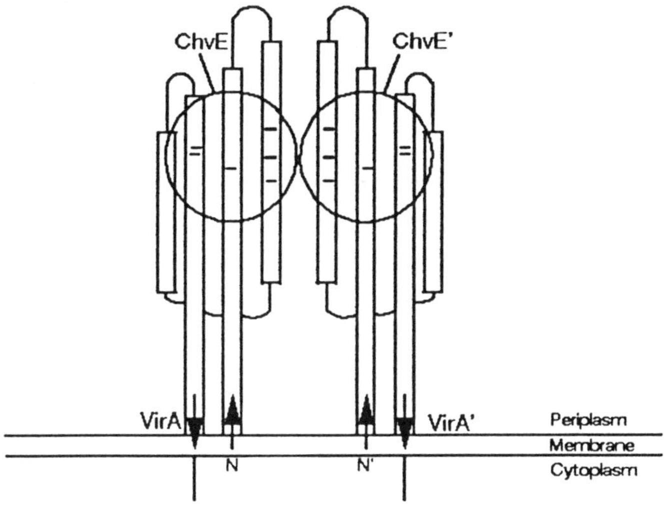

Figure 3. Model of the interaction between ChvE and VirA. Rectangles represent alpha helices. Dashes indicate locations of the six mutations in the periplasmic domain. Reprinted from with permission from Doty, *et al.* 1996 (11).

protein (5,21), the chemoreceptor Tar (23,25), and the secondary structure prediction program (IntelliGenetics) that the periplasmic domain of VirA is a four helix bundle. The relative positions of the six missense mutations are marked. Our data as well as those of others (4,31,41) suggest that ChvE recognizes a large portion of this structure rather than a single site.

The amphipathic helix of VirA may be involved in detection of phenolic compounds. Seligman and Manoil (29) found that the chemoreceptor proteins and a number of histidine kinases had a conserved putative amphipathic helix following the second transmembrane domain. They showed that the helix plays a role in stabilizing the orientation of the protein in the membrane. Figure 4 shows a helical wheel depiction of the amphipathic helix of VirA. One of our mutants had a Thr284Met change within the amphipathic

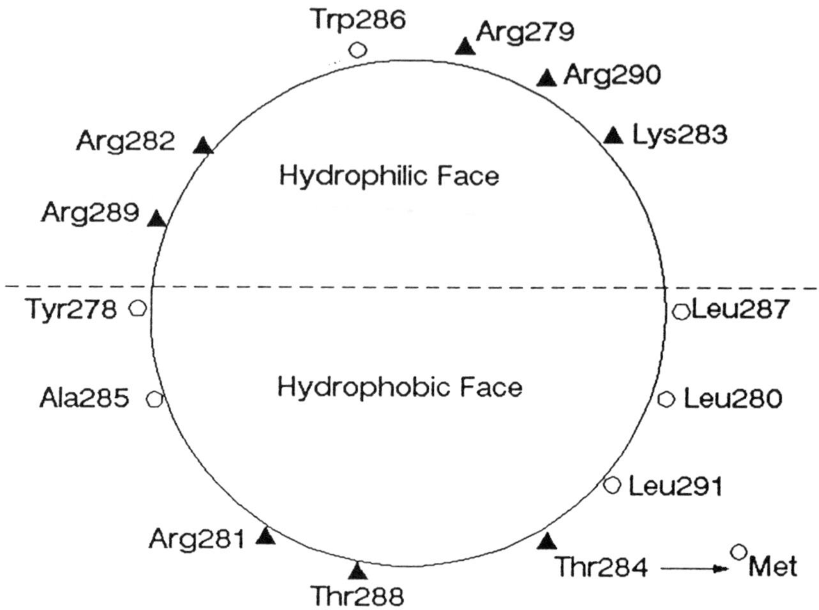

Figure 4. Helical wheel depiction of the amphipathic helix of VirA. (○) indicates a hydrophobic residue and (▲) indicates a hydrophilic residue.

helix of VirA (11). This mutation resulted in a lack of *vir* gene induction whether the mutation was in A6 *virA* or in C58 *virA* and the mutant was nearly avirulent. Since the mutant VirA protein was stable, dimerized normally, and was oriented correctly in the membrane, the phenotype of the mutant may result from the inability of the mutant protein either to detect or respond to inducing molecules. The amphipathic helix is thought to align with the inner membrane (29) which is where AS localizes (2). Perhaps VirA evolved to detect the presence of the phenolic compounds using its amphipathic helix.

Discussion

In wild-type VirA, the periplasmic domain exerts a repressive effect which can be relieved by high concentrations of AS. Deletion of the periplasmic domain results in a VirA protein which is extremely sensitive to AS even in the absence of sugar (5,13). Heath, *et al.* described a model for the activation of the VirA protein, proposing that VirA has three states of activity: Off, Standby, and On (14). ChvE with its sugar ligand interacts with VirA and releases the periplasmic domain of its repressive conformation, and places VirA into a Standby position, in which it is able to react to AS. Without sugar-bound ChvE, the few VirA molecules which randomly assume the Standby conformation could respond to AS only if AS was at a saturating concentration. Our data (11) suggest that, in the case of C58 VirA, ChvE interacts with VirA even without its sugar ligand, and that C58 VirA is unable to achieve the Standby mode without ChvE.

The transmembrane domains of chemoreceptors and sensor histidine kinases seem to be important for signal transduction within the protein and within the homodimer. Mutants of the sensor protein, EnvZ, and the chemoreceptor protein, Tar, suggest that the transmembrane domains of each monomer within a dimer may interact and that the TM1 domains of the dimer may also interact to propagate a signal (25,37,38). In an earlier mutational study of VirA, a constitutively active VirA was isolated which had a mutation in TM1, suggesting that the transmembrane domain of VirA is important for signaling (26). Also, substitution of either TM domain of VirA with that of Tar reduced *vir* gene induction (40). A *virA* mutant described by Chang and Winans (7) has a deletion of both TM domains as well as the periplasmic domain and part of the amphipathic helix which resulted in less activity and more VirA protein than wild-type. In other studies, deletion of the periplasmic domain enhanced *vir* gene induction by low levels of AS. In this mutant, though, the decrease in activity compared to that of wild-type cells suggests that the TM domains are important. The mutants which we isolated with mutations in the TM domains of both C58 and A6

VirA showed less induction by even high levels of AS, thus confirming that the transmembrane domains are important for signaling.

The proposed Standby position for VirA (14) may involve an interaction of TM1 with TM2, causing conformational shifts which enable the amphipathic helix to interact with AS. The amino-terminal tail may also be involved in the transmission of the signal by interacting with the region past the amphipathic helix. The VirA/VirG two-component system is well understood in terms of the signal molecules involved and the genes which are activated; however, it is not entirely clear how these signal molecules interact with the VirA protein.

References

1. **Alt-Moerbe, J., H. Kuhlmann, and J. Schroder**. 1989. Differences in induction of Ti plasmid virulence genes *virG* and *virD*, and continued control of *virD* expression by four external factors. Mol. Plant-Microbe Interact. **2:**301-308.

2. **Ankenbauer, R.** Unpublished data.

3. **Ankenbauer, R. G. and E. W. Nester**. 1990. Sugar-mediated induction of *Agrobacterium tumefaciens* virulence genes: structural specificity and activities of monosaccharides. J. Bacteriol. **172:**6442-6446.

4. **Banta, L. M., R. D. Joerger, V. R. Howitz, A. M. Campbell, and A. N. Binns**. 1994. Glu-255 outside the predicted ChvE binding site in VirA is crucial for sugar enhancement of acetosyringone perception by *Agrobacterium tumefaciens.*. J. Bacteriol. **176:**3242-3249.

5. **Cangelosi, G. A., R. G. Ankenbauer, and E. W. Nester**. 1990. Sugars induce the *Agrobacterium* virulence genes through a periplasmic binding protein and a transmembrane signal protein. Proc. Natl. Acad. Sci. USA **87:**6708-6712.

6. **Cangelosi, G. A., E. A. Best, G. Martinetti, and E. W. Nester**. 1991. Genetic analysis of *Agrobacterium*. Meth. Enzymology **204:**384-397.

7. **Chang, C. H. and S. C. Winans**. 1992. Functional roles assigned to the periplasmic, linker, and receiver domains of the *Agrobacterium tumefaciens* VirA protein. J. Bacteriol. **174:**7033-7039.
8. **Charles, T. C., S. L. Doty, and E. W. Nester**. 1994. Construction of *Agrobacterium* strains by electroporation of genomic DNA and its utility in analysis of chromosomal virulence mutations. Appl. Environ. Microbiol. **60:**4192-4194.
9. **Chilton, M.-D., T. Currier, S. Farrand, A. J. Bendich, M. P. Gordon, and E. W. Nester**. 1974. *Agrobacterium tumefaciens* DNA and PS8 bacteriophage DNA not detected in crown gall tumors. Proc. Natl. Acad. Sci. USA **71:**3672-3676.
10. **Doty, S. L., M. Chang, and E. W. Nester**. 1993. The chromosomal virulence gene, *chvE*, of *Agrobacterium tumefaciens* is regulated by a LysR family member. J. Bacteriol. **175:**7880-7886.
11. **Doty, S. L., M. C. Yu, J. I. Lundin, J. D. Heath, and E. W. Nester**. 1996. Mutational analysis of the sensory domain of the VirA protein of *Agrobacterium tumefaciens.*. J. Bacteriol. **178**:961-970
12. **Garfinkel, D. J. and E. W. Nester**. 1980. *Agrobacterium tumefaciens* mutants affected in crown gall tumorigenesis and octopine catabolism. J. Bacteriol. **144:**732-743.
13. **Heath, J. D.** Unpublished data.
14. **Heath, J. D., T. C. Charles, and E. W. Nester**. 1995. Ti Plasmid and Chromosomally Encoded Two-Component Systems Important in Plant Cell Transformation by *Agrobacterium* Species. *In* J.A. Hoch and T.J. Silhavy (eds.) Two-Component Signal Transduction. American Society for Microbiology, Washington, D.C.
15. **Huang, M.-L., G. A. Cangelosi, W. Halperin, and E. W. Nester**. 1990. A chromosomal *Agrobacterium tumefaciens* gene required for effective plant signal transduction. J. Bacteriol. **172:**1814-1822.
16. **Iuchi, S.** 1993. Phosphorylation/dephosphorylation of the receiver module at the conserved aspartate residue controls transphosphorylation activity of histidine kinase in sensor protein ArcB of *Escherichia coli*. J. Biol. Chem. **268:**23972-23980.
17. **Jin, S., T. Roitsch, P. J. Christie, and E. W. Nester**. 1990. The regulatory VirG protein specifically binds to a *cis*-acting regulatory

sequence involved in transcriptional activation of *Agrobacterium tumefaciens* virulence genes. J. Bacteriol. **172:**531-537.

18. **Lee, K., M. W. Dudley, K. M. Hess, D. G. Lynn, R. D. Joerger, and A. N. Binns**. 1992. Mechanism of activation of *Agrobacterium* virulence genes: identification of phenol-binding proteins. Proc. Natl. Acad. Sci. USA **89:**8666-8670.

19. **Lee, Y.-W., S. Jin, W.-S. Sim, and E. W. Nester**. 1995. Genetic evidence for direct sensing of phenolic compounds by the VirA protein of *Agrobacterium tumefaciens*. Proc. Natl. Acad. Sci. USA **92:**12245-12249.

20. **Leroux, B., M. F. Yanofsky, S. C. Winans, J. E. Ward, S. F. Ziegler, and E. W. Nester**. 1987. Characterization of the *virA* locus of *Agrobacterium tumefaciens*: a transcriptional regulator and host range determinant. EMBO **6:**849-856.

21. **Melchers, L. S., T. J. G. Regensburg-Tuink, R. B. Bourret, N. J. A. Sedee, R. A. Schilperoort, and P. J. J. Hooykaas**. 1989. Membrane topology and functional analysis of the sensory protein VirA of *Agrobacterium tumefaciens*. EMBO **8:**1919-1925.

22. **Melchers, L. S., A. J. G. Regensburg-Tuink, R. A. Schilperoort, and P. J. J. Hooykaas**. 1989. Specificity of signal molecules in the activation of *Agrobacterium* virulence gene expression. Mol. Microbiol. **3:**969-977.

23. **Milburn, M. V., G. G. Prive, D. L. Miligan, W. G. Scott, J. Yeh, J. Jancarik, D. E. J. Koshland, and S.-H. Kim**. 1991. Three-dimensional structures of the ligand-binding domain of a transmembrane receptor with and without a ligand: the aspartate receptor of bacterial chemotaxis. Science **254:**1342-1347.

24. **Ninfa, E. G., M. R. Atkinson, E. S. Kamberov, and A. J. Ninfa**. 1993. Mechanism of autophosphorylation of *Escherichia coli* nitrogen regulator II (NRII or NtrB): *trans*-phosphorylation between subunits. J. Bacteriol. **175:**7024-7032.

25. **Parkinson, J. S.** 1993. Signal transduction schemes of bacteria. Cell **73:**857-871.

26. **Pazour, G. J., C. N. Ta, and A. Das**. 1991. Mutants of *Agrobacterium tumefaciens* with elevated *vir* gene expression. Proc. Natl. Acad. Sci. USA **88:**6941-6945.

27. **Raineri, D. M.** Unpublished data.

28. **Raineri, D. M., M. I. Boulton, J. W. Davies, and E. W. Nester**. 1993. VirA, the plant-signal receptor, is responsible for the Ti plasmid-specific transfer of DNA to maize by *Agrobacterium*. Proc. Natl. Acad. Sci. USA **90:**3549-3553.

29. **Seligman, L.** 1995. Ph.D. thesis. University of Washington.

30. **Seligman, L. and C. Manoil**. 1994. An amphipathic sequence determinant of membrane protein topology. J. Biol. Chem. **269:**19888-19896.

31. **Shimoda, N., A. Toyoda-Yamamoto, S. Aoki, and Y. Machida**. 1993. Genetic evidence for an interaction between the VirA sensor protein and the ChvE sugar-binding protein of *Agrobacterium*. J. Biol. Chem. **268:**26552-26558.

32. **Shimoda, N., A. Toyoda-Yamamoto, J. Nagamine, S. Usami, M. Katayama, Y. Sakagami, and Y. Machida**. 1990. Control of expression of *Agrobacterium vir* genes by synergistic actions of phenolic signal molecules and monosaccharides. Proc. Natl. Acad. Sci. USA **87:**6684-6688.

33. **Stachel, S. E., G. An, C. Flores, and E. W. Nester**. 1985. A Tn*3* *lacZ* transposon for the random generation of β-galactosidase gene fusions: application to the analysis of gene expression in *Agrobacterium*. EMBO **4:**891-898.

34. **Stachel, S. E., E. W. Nester, and P. C. Zambryski**. 1986. A plant cell factor induces *Agrobacterium tumefaciens vir* gene expression. Nature **318:**624-629.

35. **Stock, J. B., M. G. Surette, M. Levit, and P. Park**. 1995. Two-Component Signal Transduction Systems: Structure-Function Relationships and Mechanisms of Catalysis. *In* J.A. Hoch and T.J. Silhavy, (eds.) Two-Component Signal Transduction. ASM Press, Washington, D.C.

36. **Swanson, R. V., R. B. Bourret, and M. I. Simon**. 1993. Intermolecular complementation of the kinase activity of CheA. Mol. Microbiol. **8:**435-441.

37. **Tokishita, S., A. Kojima, and T. Mizuno**. 1992. Transmembrane signal transduction and osmoregulation in *Escherichia coli*: functional importance of the transmembrane regions of membrane-located protein kinase, EnvZ. J. Biochem. **111:**707-713.

38. **Tokishita, S. and T. Mizuno**. 1994. Transmembrane signal transduction by the *Escherichia coli* osmotic sensor, EnvZ: intermolecular complementation of transmembrane signaling. Mol. Microbiol. **13:**435-444.

39. **Turk, S. C. H. J., L. S. Melchers, H. D. Dulk-Ras, A. J. G. Regensburg-Tuink, and P. J. J. Hooykaas**. 1991. Environmental conditions differentially affect *vir* gene induction in different *Agrobacterium* strains. Role of the VirA sensor protein. Plant Mol. Biol. **16:**1051-1059.

40. **Turk, S. C. H. J., R. P. van Lange, T. J. G. Regensburg-Tuink, and P. J. J. Hooykaas**. 1994. Localization of the VirA domain involved in acetosyringone-mediated *vir* gene induction in *Agrobacterium tumefaciens*. Plant Mol. Biol. **25:**899-907.

41. **Turk, S. C. H. J., R. P. van Lange, E. Sonneveld, and P. J. J. Hooykaas**. 1993. The chimeric VirA-Tar receptor protein is locked into a highly responsive state. J. Bacteriol. **175:**5706-5709.

42. **Veluthambi, K., M. Krishnan, J. H. Gould, R. H. Smith, and S. B. Gelvin**. 1989. Opines stimulate induction of the *vir* genes of the *Agrobacterium tumefaciens* Ti plasmid. J. Bacteriol. **171:**3696-3703.

43. **Winans, S. C., R. A. Kerstetter, and E. W. Nester**. 1988. Transcriptional regulation of the *virA* and *virG* genes of *Agrobacterium tumefaciens*.. J. Bacteriol. **170:**4047-4054.

44. **Wolfe, A. J. and R. C. Stewart**. 1993. The short form of the CheA protein restores kinase activity and chemotactic ability to kinase-deficient mutants. Proc. Natl. Acad. Sci. USA **90:**1518-1522.

45. **Yang, Y. and M. Inouye**. 1992. Intermolecular complementation between two defective mutant signal-transducing receptors. Proc. Natl. Acad. Sci. USA **88:**11057-11061.

CHAPTER 2

Environmental and Genetic Factors Affecting RSF1010 Mobilization Between Strains of *Agrobacterium tumefaciens*

Karla Jean Fullner and Eugene W. Nester

Department of Microbiology, Box 357242
University of Washington
Seattle, Washington 98195-7242 USA

ABSTRACT: A quantitative assay to monitor function of the T-DNA transfer machinery in oncogenic strains of *Agrobacterium tumefaciens* was developed. This assay is based on the observation that the IncQ plasmid RSF1010 is transferred to other strains of *A. tumefaciens* using the same machinery as the T-DNA uses to enter plant cells. Transfer of a conjugative plasmid between bacteria is ideal for the study of the T-DNA transfer machinery since factors affecting T-DNA processing or activities within the plant do not impact the assay. A protocol for RSF1010 transfer was established and important parameters were optimized including temperature, donor-to-recipient ratios, and induction protocols. Strain A348 carrying transposon insertions or deletions in the *vir* region were then tested for function in the RSF1010 transfer assay. Strains with mutations in the *virA*, *virG*, *virB1-11* or *virD4* were unable to transfer a derivative of RSF1010, pML122, to a recipient strain of *A. tumefaciens*. Mutation of other *vir* genes did not abolish transfer. Strains with mutations affecting attachment to plant cells also did not reduce transfer.

Introduction

Genetic transformation of plant cells by *Agrobacterium* is a widely used tool in plant molecular biology. Development of *Agrobacterium*-based transformation has yielded many transgenic plants for use either in the study of plant molecular biology or in the improvement of economically important crops. The usefulness of this system has expanded despite a limited understanding of the process by which *Agrobacterium* introduces DNA into plant cells. It is clear that the intermediate in transfer is the T-strand, which is composed of the single-stranded T-DNA covalently attached to the endonucleolytic enzyme VirD2 (1, 21, 23). This complex is transferred to the plant dependent upon the gene products of the *virB* and *virD* operons by a mechanism which most likely resembles bacterial conjugation (18). However, a clear picture of the transfer process is not yet at hand.

This lack of understanding is due in part to the absence of an easily performed, inexpensive assay for monitoring function of the T-DNA transfer machinery. Although mutants in the *virB* and *virD* operons are avirulent, tumorigenesis is a poor assay for transfer since tumor induction involves multiple steps. Agroinfection is another assay which has been used to study transfer of T-DNA (15, 20). For this assay, viral genomes are cloned in a tandem head-to-tail arrangement between T-DNA borders. Following inoculation of susceptible plants, viral symptoms develop only if the virus DNA is transferred to the plant cells. Although agroinfection does not require stable integration of the DNA into the plant genome, this assay does require targeting of the T-DNA to the plant nucleus for viral genes to be expressed. A further complication of agroinfection assays is the requirement of special licenses and a large number of plants. A PCR-based assay has also been developed for monitoring T-DNA transfer (24). Following infection of tobacco BY-2 suspension cells by *A. tumefaciens*, the bacteria are washed off the plant cells and PCR is used to detect T-DNA intermediates within the plant cell cytoplasm. This assay is sensitive, quantitative, and specific for T-DNA transfer, but the assay is not amenable to large-scale screening protocols.

Function of the T-DNA transfer machinery has also been measured by examining mobilization of the plasmid RSF1010 between strains of *A. tumefaciens* (2, 14). RSF1010 is a plasmid of the IncQ group which is non-self conjugal in that it does not carry all the genetic information to enable its conjugation into other bacteria. However, RSF1010 can be transferred to other bacteria by utilizing the transfer functions of another conjugative plasmid from a different incompatibility group resident within the same bacterial cell (22). Interestingly, *A. tumefaciens* can transfer RSF1010 into plant cells and into other strains of *A. tumefaciens* without the assistance of a co-resident plasmid (2, 5). Transfer of RSF1010 from *A. tumefaciens* requires genes within the *virB* and *virD* operons suggesting that RSF1010 uses the T-DNA transfer machinery for its entry into plant cells and agrobacteria (2, 6). Monitoring conjugal transfer of RSF1010 between bacteria is a convenient system for the study of the T-DNA transfer machinery since the transfer event is independent of T-DNA processing and subsequent events occurring within the plant cell. Conjugation assays are also amenable to inexpensive large-scale screening to identify point mutations in proteins essential for transfer or to isolate suppressors of previously generated point mutants. However, these wider applications require improvement of the RSF1010 transfer efficiencies from strains with intact T-DNA regions in order to correlate RSF1010 conjugation results with tumorigenesis.

In this study, the mobilization of RSF1010 from the oncogenic strain A348 is characterized. Several parameters were optimized which significantly improve the overall transfer efficiency. A mutant analysis showed that mobilization of RSF1010 from A348 requires the *virA*, *virG*, *virB*, and *virD4* genes but does not require other *vir* genes or chromosomal attachment genes.

Materials and Methods

For conjugation assays, *A. tumefaciens* strain A348 [C58 chromosome with pTiA6, Rif^R (19)] was used as a donor of the IncQ group plasmid pML122 [Km^R, Gm^R (17)]. All donor strains with mutations in *vir* regions are derivatives of A348 (3, 8, 13, 14). ME42 and ME69 (9) were cured of pPH1JI as previously described (14) to create the donor strains At12512 and At12513, respectively. Strain UIA143 [C58 chromosome, no Ti plasmid, Ery ^{R}(10)] was used as the recipient. Conjugation assays were performed as follows, except as noted in text. Cultures of donor and recipient strains were grown overnight in MG/L with antibiotics at 28°C. Cultures were diluted into CIB broth [1X CIB Salts (14), 2% glucose, 62.5 mM potassium phosphate pH 5.3, and 200 μM acetosyringone] without antibiotics to an OD_{600} of 0.25 and were incubated at 28°C with shaking for 6 hr. Donor and recipients cells were mixed together at a ratio of 7 to 1 in a 1.5 ml microfuge tube to a combined OD_{600} of 0.25. 5 μl of the mating mix was spotted in the center of a 35 mm petri dish containing 4 ml of CIB agar. Plates were inverted and incubated at 19°C for three days. Bacteria were recovered by washing the surface of the agar 3 times with 0.9% NaCl to yield a pooled volume of 1 ml. Recovered bacteria were plated onto LB agar supplemented with the appropriate antibiotics to select for transconjugants. Plates were incubated at 28°C for 4-5 days and colonies counted.

Results and Discussion

Temperature affects the transfer of pML122 between strains of *Agrobacterium*. In the course of performing assays on mobilization of the IncQ group plasmid pML122, we noted that replicates within the same experiment gave reproducible transfer frequencies; but, the frequencies from the same donor strain varied as much as 100-fold between experiments performed on different days. At times, wild-type strain A348 failed to transfer entirely. These fluctuations occurred even when the same batch of CIB mating agar plates was used. These observations suggested that an uncontrolled environmental factor was affecting transfer.

The temperature within the incubator used for conjugation experiments was monitored over a week-long period, and we found that the temperature fluctuated between 25°C and 29°C. To examine whether small temperature changes could affect the conjugation frequency from A348, the same mixture of donor

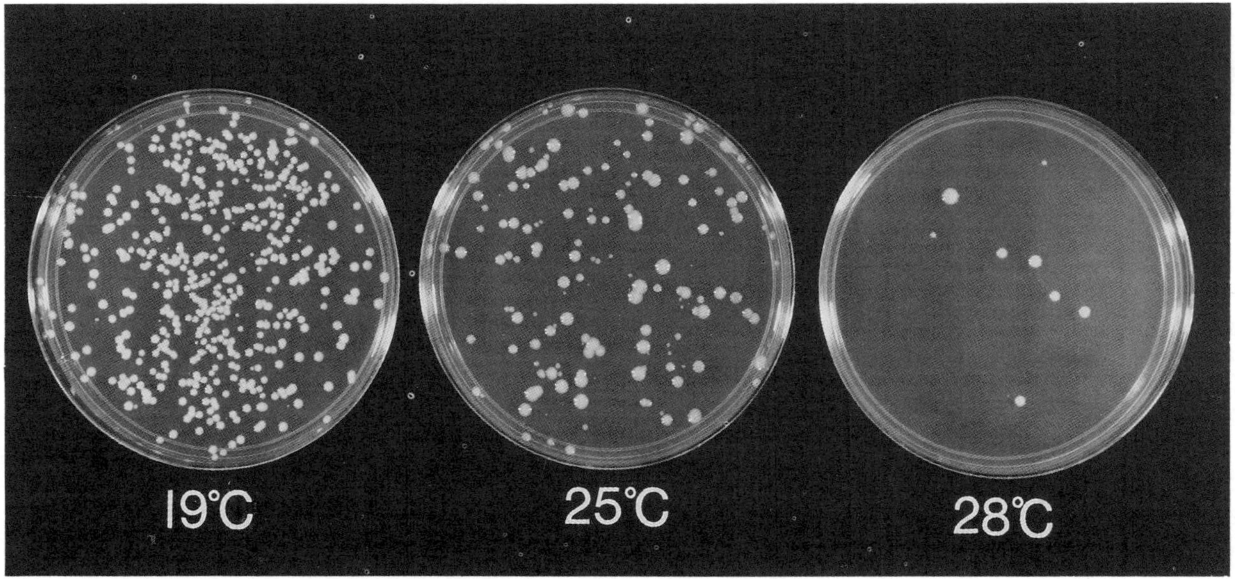

Figure 1. Recovery of transconjugants from matings of A348(pML122) with UIA143 at various temperatures. Transconjugants were recovered in 1 ml 0.9% NaCl and 0.1 ml was plated for 19°C and 25°C and 1.0 ml was plated for 28°C. The transfer frequencies (transconjugants/input donor) corresponding to the plates shown are: 19°C, 3 x 10^{-3}; 25°C, 5 x 10^{-4}; and 28°C, 2 x 10^{-6}.

A348(pML122) cells and recipient UIA143 cells was spotted onto mating plates which were then incubated at 19°C, 25°C, and 28°C. Over the three day mating period, the temperature inside all three incubators did not fluctuate more than 0.5°C. The number of transconjugants recovered from matings incubated at 19°C was markedly greater than when matings were incubated at higher temperatures (Fig. 1). Further experiments demonstrated that 19°C is the optimal temperature for function of the T-DNA transfer machinery in conjugative transfer (13).

Donor-to-recipient ratios. The ratio of input donors to recipients is known to affect the overall frequency of conjugation in numerous systems. After mating A348(pML122) with UIA143, recovery of transconjugants was optimal when an input donor-to-recipient ratio of 7 to 1 was used (Table 1). In this experiment, we noted that although the number of transconjugants

Table 1. Effect of varying donor-to-recipient ratios on the mobilization of RSF1010 by *A. tumefaciens*

		Frequency of transfer of pML122	
Ratio (donor: recipients)[a]	Recovered transconjugants	Transconjugants/ input donor	Transconjugants / recovered recipient
117:1	950	3.7×10^{-4}	1.4×10^{-5}
10:1	2230	9.5×10^{-4}	5.3×10^{-6}
7:1	2480	1.1×10^{-3}	6.5×10^{-6}
3:1	2250	1.3×10^{-3}	2.9×10^{-6}
1:1	1420	1.4×10^{-3}	2.4×10^{-6}
1:2	1100	1.4×10^{-3}	1.5×10^{-6}
1:5	465	1.6×10^{-3}	7.2×10^{-6}

[a]For all assays, the total number of cells used was constant at 2.5×10^{6} bacteria per assay.

recovered varied, the frequency of transfer per input donor did not except at the extreme ratio of 117:1. Based on these results, all future experiments were prepared at a ratio of 7 to 1 and the transfer frequencies were calculated by the per input donor method.

Mobilization of pML122 from A348 requires induction of *vir* genes. To establish that transfer of pML122 from the oncogenic strain A348 was dependent on *vir* genes, transfer in the presence and absence of the *vir* gene inducer acetosyringone (AS) was tested. Following mating of A348(pML122) with UIA143, transconjugants were recovered only when AS was present in the CIB agar showing that induction of *vir* genes was essential for the mobilization of RSF1010 from strain A348 (Table 2). Addition of AS to the culture broth during the pre-induction step did not

Table 2. Mobilization of pML122 from *A. tumefaciens* under a variety of inducing conditions[a]

Pre-treatment	Mating agar	Conjugation frequency (x 10^{-4})
CIB + AS	CIB + AS	17 ± 5
CIB + AS	CIB - AS	<0.007
CIB - AS	CIB + AS	14 ± 7
CIB - AS	CIB - AS	<0.006

[a]Bacteria collected from overnight cultures of A348(pML122) and UIA143 cells were resuspended in the indicated broth, and incubated with shaking for 6 hours at room temperature. The cells were mixed and were spotted on the indicated mating agar. Conjugation frequencies were calculated as transconjugants per input donor and represent the average of triplicates from a single experiment.

affect the conjugation frequency (Table 2) even though *vir* genes were induced in this time period (data not shown). Further, induction of *vir* genes by AS prior to mating did not permit mating on agar medium which did not contain AS (Table 2). These results suggested that pre-induction was not an essential step in the conjugation protocol. However, cells grown for 6 hours in the rich medium MG/L served as poor donors of pML122 (data not shown). Presumably, pre-induction is essential because the bacteria are acclimatized to the minimal induction medium prior to initiation of mating. These results further suggest that structures essential for transfer are either rapidly turned over such that the *vir* genes must be stimulated over the entire mating period or that the structures important for transfer are assembled only on solid substrates and, thus, are not present in the broth cultures.

Mobilization of pML122 requires *virA*, *virG*, *virB1-11* and *virD4*. To establish which *vir* genes are essential for transfer of pML122 from A348, strains with deletions or insertions in *virA-H* and *chvE* were tested for their ability to serve as donors of ML122 in a standard conjugation assay. Donor strains with a mutation in any of the three regulatory genes, *virA*, *virG*, or *chvE*, were unable to mobilize pML122 (Table 3). These data confirm that *vir* gene induction is essential for IncQ plasmid transfer from oncogenic strains. In addition, donor strains with mutations in the *virB* and *virD* operons also did not transfer pML122 (Table 3). These data are consistent with results from agroinfection and PCR-based studies which showed that the *virB* and *virD* operons are essential for transfer of T-DNA to plant cells demonstrating that RSF1010 uses the T-DNA transfer machinery for its movement between bacteria (16, 24). Other *vir* genes, *virC*, *virE2*, *virF*, and *virH*, were not essential for transfer although there were differences in the observed transfer frequency (Table 3). In all, *virA*, *virG*, *virB*, and *virD* are the only known *vir* operons essential for mobilization of RSF1010.

Since the *virB* and *virD* operons contain multiple genes essential for virulence, strains with nonpolar mutations in these operons were tested for their ability to donate pML122. These studies showed that *virB2-11* were absolutely essential for mobilization (12). However, a deletion of the *virB1* gene functioned in 3 of 15 assays but at a frequency 1000-fold below wild-type (12). Analysis of nonpolar mutations within the *virD* operon showed that the requirement for the *virD* operon is limited to *virD4* (12). Thus, we conclude that *virB1-11* and *virD4* are each required for efficient mobilization of RSF1010 between strains of *A. tumefaciens*.

Table 3. Mobilization of pML122 from mutant backgrounds of *A. tumefaciens*

Donor Strain[a]	Site of mutation[b]	Conjugation frequency (x 10^{-6})[c]
A348	Wild-type	350
At11068	*virA*	<1
At12508	*virG*	<1
At12515	*chvE*	<1
At11067	*virB1*	<2
At12511	*virD1*	<1
At12506	*virD4*	<1
At12509	*virC*	1100
At12516	*virE2*	140
At12510	*virF*	160
At12507	*virH*	190

[a]All donor strains carry pML122.

[b]Mutations are either deletions or transposon insertions which are polar on downstream genes.

[c]Conjugation frequencies were calculated as transconjugants per input donor and represent averages of triplicates from a single assay.

***virB* and *virD4* are sufficient for RSF1010 mobilization.** To establish whether any other genes on the Ti plasmid are essential for functional transfer, a minimal transfer plasmid comprised of *virA*, *virG*, and *virB*, expressed from their native promoters, and *virD4*, expressed from the *virB* promoter was constructed (11). The Ti plasmidless strain A136 mobilized the IncQ plasmid pML122Km when it carried this minimal transfer plasmid but not when it carried a similar plasmid which carried only *virA* and *virG* (11). These data demonstrate that no other Ti plasmid genes are essential for RSF1010 mobilization and suggest that *virB* and *virD4* encode the complete T-DNA transfer machinery.

The role of attachment in RSF1010 mobilization. In addition to genes found on the Ti plasmid, three chromosomal genes were tested for their contribution to RSF1010 mobilization. Strains with transposon insertions in *chvA*, *chvB*, and *exoC* are avirulent due to the absence of cyclic-beta-1,2-glucan which either directly or indirectly affects the ability of these mutants to attach to plant cells (4). The *chvB* mutant At12513 transferred pML122 at a frequency comparable to wild-type (10^{-3}) demonstrating that the absence of cyclic-beta-1,2-glucan does not affect the ability of the transfer machinery to function. Similarly, the *chvA* mutant At12512, which cannot export cyclic-beta-1,2-glucan, transferred pML122 although the frequency of transfer was erratic, varying between 10-fold below and 10-fold above wild-type frequencies in different experiments (10^{-2} to 10^{-4}). The *exoC* mutant A5129 (7) consistently transferred pML122 at a frequency 10-fold greater than A348 (10^{-2}). It is possible that the absence of an exopolysaccharide capsule in the *exoC* mutant strain favored increased contact between the bacterial strains leading to the higher frequency of transfer. In all, these results show that mutations in *chvA*, *chvB*, and *exoC* are primarily defective in attachment to plant cells and do not carry pleiotrophic phenotypes which disrupt proper assembly of the transport complex.

Future applications of the conjugation assay. The RSF1010 conjugation may be useful to study specific interactions between the eleven VirB proteins and VirD4. The high frequency of transfer allows one to identify point mutations which lead to loss of function in the transfer machinery. Pilot studies have shown that donors carrying point mutations in the *virB4* gene (14) could be distinguished from the wild-type donors in a two-step plate patching assay. For the first step, single bacterial colonies, either wild-type or mutant, were applied with a toothpick to a 100 mm CIB agar plate spread with recipient and the bacteria were allowed to mate for four days. Transconjugants were identified by replica plating the matings onto selective media. Using this protocol, wild-type and mutant cells were clearly discernible. No false positive or false negative transconjugants arose provided the bacterial cells used for mating were from freshly grown plates. Using this patching assay, thousands of potential transfer machinery mutants could be quickly screened. Since mutations can be detected in oncogenic strains, mutants identified by RSF1010 transfer can be subsequently screened by the more sensitive tumor induction assay.

After transfer mutants are identified, RSF1010 mobilization could again be used to screen for second site suppressors to identify whether the transfer protein is involved in a protein-protein interaction. In the design of such protocols, however, it is important to note that IncW, IncP, and IncQ plasmids are poorly maintained within the same cell even in the presence of antibiotics (data not shown). This problem can be easily circumvented. The mutant transfer gene would be carried on an IncP or IncW plasmid along with *virA* and *virG* in a Ti plasmidless background. The remaining transfer genes could be cloned into the polylinker of the IncQ plasmid pML122. The transfer genes carried on pML122 could then be mutated, either chemically or by passage through a mutator strain, and the resultant mutant plasmids moved into the A136(IncP or W) strain. Individual transformants could then be screened by assaying for transfer of the IncQ plasmid into the recipient. As the transferred IncQ plasmid carries the suppressing mutation, the mutant DNA sequences can be easily recovered from the recipient strain.

The quantitative nature of the RSF1010 conjugation assay also allows for assaying subtle changes in function of the T-DNA transfer machinery which may not be detectable by tumorigenesis. This application of RSF1010 mobilization has already been demonstrated in a study on the protein VirB4 (14). Strains merodiploid for VirB4 showed decreased frequencies of RSF1010 transfer, demonstrating that mutations in the mononucleotide-binding domain are dominant (14). This result indicated that VirB4 functions within a complex. Using the same strains, suppression of transfer was not seen by tumorigenesis. Thus, assessment of transfer events using RSF1010 mobilization assays can lead to different conclusions regarding the function of the T-DNA transfer machinery.

Summary. Our studies show that RSF1010 mobilization from wild-type A348 requires the same machinery as T-DNA transfer to plant cells. We have optimized the transfer protocol leading to a significant increase in the transfer frequency compared to previously reported values (2, 14). These improvements make RSF1010 mobilization useful as a tool to study function of the individual components of the T-DNA transfer machinery, VirB1-11 and VirD4. This assay should also have applications beyond those mentioned since the assay is easy, inexpensive, and amenable to large-scale screening for mutants. These factors make RSF1010 mobilization a very useful system for the study of transfer from *Agrobacterium*.

References

1. **Albright, L. M., M. F. Yanofsky, B. Leroux, D. Q. Ma and E. W. Nester.** 1987. Processing of the T-DNA of *Agrobacterium tumefaciens* generates border nicks and linear, single-stranded T-DNA. J. Bacteriol. **169**:1046-1055.

2. **Beijersbergen, A., A. Dulk-Ras, R. Schilperoort and P. Hooykaas.** 1992. Conjugative transfer by the virulence system of *Agrobacterium tumefaciens*. Science **256**:1324-1326.

3. **Berger, B. R. and P. J. Christie.** 1994. Genetic complementation analysis of the *Agrobacterium tumefaciens virB* operon: *virB2* through *virB11* are essential virulence genes. J. Bacteriol. **176**:3646-3660.
4. **Binns, A. N. and M. F. Thomashow.** 1988. Cell biology of *Agrobacterium* infection and transformation of plants. Annu. Rev. Microbiol. **42**:575-606.
5. **Buchanan-Wollaston, V., J. Passiatore and F. Cannon.** 1987. The *mob* and *oriT* mobilization functions of a bacterial plasmid promote its transfer to plants. Nature **328**:172-175.
6. **Buchanan-Wollaston, V., J. E. Passiatore and F. Cannon.** 1988. The effect of *vir* mutations on plasmid transfer into plants, p. 281-282. *In* R. Palcios and D.P.S. Verma (eds.), Molecular genetics of plant-microbe interactions. APS Press, St. Paul, MN.
7. **Cangelosi, G. A., L. Hung, V. Puvanesarajah, G. Stacey, D. A. Ozga, J. A. Leigh and E. W. Nester.** 1987. Common loci for *Agrobacterium tumefaciens* and *Rhizobium meliloti* exopolysaccharide synthesis and their roles in plant interactions. J. Bacteriol. **169**:2086-2091.
8. **Dale, E., A. Binns and J. Ward.** 1993. Construction and characterization of Tn*5virB*, a transposon that generates nonpolar mutations, and its use to define *virB8* as an essential virulence gene in *Agrobacterium tumefaciens*. J. Bacteriol. **175**:887-891.
9. **Douglas, C. J., R. J. Staneloni, R. A. Rubin and E. W. Nester.** 1985. Identification and genetic analysis of an *Agrobacterium tumefaciens* chromosomal virulence region. J. Bacteriol. **161**:850-860.
10. **Farrand, S. K., S. P. O'Morchoe and J. McCutchan.** 1989. Construction of an *Agrobacterium tumefaciens* C58 *recA* mutant. J. Bacteriol. **171**:5314-5321.
11. **Fullner, K. J.** Unpublished results.
12. **Fullner, K. J. and E. W. Nester.** Unpublished results.
13. **Fullner, K. J. and E. W. Nester.** 1996. Temperature affects the T-DNA transfer machinery of *Agrobacterium tumefaciens*. J. Bacteriol. **178**:1498-1504.
14. **Fullner, K. J., K. M. Stephens and E. W. Nester.** 1994. An essential virulence protein of *Agrobacterium tumefaciens*, VirB4,

requires an intact mononucleotide binding domain to function in transfer of T-DNA. Mol. Gen. Genet. **245**:705-715.

15. **Grimsley, N., B. Hohn, T. Hohn and R. Walden.** 1986. "Agroinfection", an alternative route for viral infection of plants by using the Ti-plasmid. Proc. Natl. Acad. Sci. USA **83**:3282-3286.

16. **Grimsley, N., B. Hohn, C. Ramos, C. Kado and P. Rogowsky.** 1989. DNA transfer from *Agrobacterium* to *Zea mays* or *Brassica* by agroinfection is dependent on bacterial virulence functions. Mol. Gen. Genet. **217**:309-316.

17. **Labes, M., A. Pühler and R. Simon.** 1990. A new family of RSF1010-derived expression and *lac*-fusion broad-host-range vectors for gram-negative bacteria. Gene **89**:37-46.

18. **Lessl, M. and E. Lanka.** 1994. Common mechanisms in bacterial conjugation and Ti-mediated T-DNA transfer to plant cells. Cell 77:321-324.

19. **Sciaky, D., A. L. Montoya and M. D. Chilton.** 1978. Fingerprints of *Agrobacterium* Ti plasmids. Plasmid **1**:238-253.

20. **Shirasu, K., Z. Koukolikova-Nicola, B. Hohn and C. Kado.** 1994. An inner-membrane-associated virulence protein essential for T-DNA transfer from *Agrobacterium tumefaciens* to plants exhibits ATPase activity and similarities to conjugative transfer genes. Mol. Microbiol. **11**:581-588.

21. **Stachel, S. E., B. Timmerman and P. Zambryski.** 1987. Activation of *Agrobacterium tumefaciens vir* gene expression generates multiple single-stranded T-strand molecules from the pTiA6 T-region: requirement for 5' *virD* gene products. EMBO J. **6**:857-863.

22. **Willetts, N. and C. Crowther.** 1981. Mobilization of the non-conjugative IncQ plasmid RSF1010. Genet. Res., Camb. **37**:311-316.

23. **Young, C. and E. W. Nester.** 1988. Association of the VirD2 protein with the 5' end of T-strands in *Agrobacterium tumefaciens*. J. Bacteriol. **170**:3367-3374.

24. **Yusibov, V. M., T. R. Steck, V. Gupta and S. B. Gelvin.** 1994. Association of single-stranded transferred DNA from

Agrobacterium tumefaciens with tobacco cells. Proc. Natl. Acad. Sci. USA **91**:2994-2998.

CHAPTER 3

The role of virulence proteins in nuclear targeting and integration of T-DNA

L. Rossi, A.M. Bravo Angel, B. Tinland* and B. Hohn

Friedrich Miescher-Institut, 4002 Basel, Switzerland.
Telephone: (061) 697-6680 / Telefax: (061) 697-3976

* Present address: ETH-Zentrum, CH 8092 Zürich, Switzerland

Introduction

Most probably DNA transfer to plants has evolved from bacterial conjugation (1). There is conservation of bacterial proteins interacting with the transferred DNAs; strong similarities in sequence and in operon-structure of proteins forming transduction channels have been found; there is even conservation of the sequence at which the transferred DNAs are cleaved at initiation of transport; in both cases the origin-specific endonucleases bind to the 5' terminus of the cleaved DNA; in the analyzed cases single-stranded versions of plasmid-or T-DNA are transported to the recipient cell. However, during this evolution adaptations were necessary to account for the very different cellular and molecular environments the transferred DNA encountered in the eukaryotic recipient. The cellular recognition of the target cell led to the wounded-plant-cell bacterium-cell signalling as it is known today. Establishment of the actual contact and transfer is a difficult process to study but the plant cell nature of the recipient certainly required changes in strategies. In addition, a plant cell contains organelles such as nuclei, chloroplasts, mitochondria or vacuoles which have to be specifically targeted or avoided. Finally, the transported DNA has to find a stable place in the nuclear DNA by integrating into it - a mechanism surely different from plasmid recircularisation or homology-driven integration into the DNA of the bacterial recipient.

In this review we will concentrate on nuclear entrance and integration of T-DNA and summarize recent data describing the roles virulence proteins play in these processes.

Nuclear targeting of the T-DNA. So far no cases have been reported in which uncomplexed nucleic acids traverse the cytoplasmic/nuclear barrier and establish themselves in the nucleus. Animal viruses such as SV40 invade nuclei as particles (2), retroviruses as preintegration complexes (3), small nuclear RNAs as snRNPs (4) whereas the single-stranded plant geminiviral DNA is transported to the nucleus most likely complexed to a protein (5).

Also the T-DNA is transferred to the plant cell nucleus as a ss-DNA protein complex (6, 7). The proteins which are thought to compose this complex, the VirD2 and VirE2 proteins, are encoded in the virulence region of the Ti-plasmid. VirD2 is a site-specific endonuclease which processes the T-DNA and attaches covalently to the 5' end of the released single-stranded T-DNA (ss-T-DNA), whereas the VirE2 protein binds to single-stranded DNA unspecifically and in an highly cooperative manner *in vitro* (8-13). VirE2 was shown to be required for efficient transfer of the T-DNA (9, 14-17). Whereas its presence is not required for the bacterial steps of the transfer (ss-T-DNA processing and export from the *A. tumefaciens* cell), the presence VirE2 in the plant is necessary for efficient T-DNA transfer (18). The VirD2 and VirE2 proteins were found to contain nuclear localization signals (NLSs): VirD2 contains an N-terminal monopartite NLS and a C-terminal bipartite NLS and VirE2 contains two bipartite NLSs. These NLS sequences, when fused to reporter proteins, were able to render them nuclear (18-20). Analysis of the relative strength of these NLSs for nuclear targeting revealed that VirD2 fusions to the beta-glucuronidase protein localized more efficiently into the nucleus than the VirE2 fusions (21). In maize and tobacco leaf tissue the protein fusion carrying the VirD2 protein efficiently localized to the nucleus, whereas only 70% of the VirE2 fusions localized to the nucleus. Surprisingly however, none of the protein fusions localized to the nucleus of mature roots of maize and tobacco. Thus, there seems to be a developmental component to nuclear uptake of at least the virulence proteins VirD2 and VirE2.

The NLS sequences of VirE2 and VirD2 were then studied in the context of T-DNA transfer. The N-terminal NLS of VirD2 was found to be unimportant for T-DNA transfer (22) whereas the C-terminal NLS was found to be essential (22-24). Precise deletion of this latter NLS reduced T-DNA transfer efficiency to about 5% compared to the wild type (23). This residual activity could be due to passive import of the complex into the nucleus during disintegration of the nuclear membrane upon cell division [as described for the nuclear import of the HIV-1 preintegration complex (25)] or to cryptic NLS sequences in VirD2. Alternatively, it could be due to the nuclear targeting properties of VirE2 molecules covering the T-DNA. This last possibility was tested. It was found that the VirE2 protein does not account for the residual transfer activity of the *virD2* C-terminal NLS mutant; in the absence of VirE2, the *virD2* NLS mutant resulted in a T-DNA transfer efficiency of about 7% compared to the wild-type VirD2-containing, VirE2-protein-deficient strain (17).

Out of the four NLSs present on VirE2 and VirD2, only the bipartite NLS of VirD2 seems to be important for T-DNA transfer. Therefore, it is the prime candidate for targeting the T-complex to the nucleus. Results obtained studying the significance of the NLSs of VirD2 and VirE2 in their biological context are apparently contradicting in part the results obtained in experiments using fusion strategies. It has been observed that NLSs and DNA-or RNA-binding motifs directly overlap or are immediately adjacent to each other in more than 79% of the nuclear DNA or RNA binding proteins with proven NLS sequences (26). For transcription factors or other DNA binding proteins these two functions are not mutually exclusive since first the protein is targeted to the nucleus and then it binds to DNA. The binding to the DNA may aid in nuclear retention of these proteins. In contrast, in order to be able to target the T-DNA to the nucleus, the virulence proteins must first bind the T-DNA and then target the complex to the nucleus. Binding to the T-DNA may mask the region acting as NLS on free proteins, thus preventing these sequences from acting as nuclear targeting sequences for the T-complex. Indeed, deletion or point mutation analysis of the NLSs of VirE2 revealed that also the cooperative DNA binding properties of the protein were affected (18, 21). Thus, probably the NLS regions of the protein are involved in intermolecular interactions (protein-protein and protein-DNA), hiding them from the surface of the protein. The N-terminal NLS of VirD2 is only 10 amino acids away from the tyrosine which is linked to the T-DNA. The T-DNA may thus mask the N-terminal NLS either directly, or indirectly by modifying the conformation of the VirD2 protein bound to it. In contrast, the C-terminal NLS of VirD2 covers a region of the protein which is not necessary for processing, the first step of T-DNA transfer (Bravo Angel, AM; unpublished results), and it is thus not involved in protein-DNA interactions. This region of the protein may be exposed at the surface of the complex and interact with the plant NLS receptor which would initiate nuclear import of the T-complex. The residual T-DNA transfer activity measured using the *virD2* NLS mutant could be due, as described above, to a cryptic NLS present in VirD2 or to entrance of the T-DNA into the nucleus during cell division.

Could VirE2 play an important role for T-DNA transfer in an NLS independent way? The nuclear import mechanism can be generally separated into two distinct steps. The first is an NLS dependent binding of the nuclear import substrate to the cytoplasmic side of the nuclear pore complex. The second is an ATP dependent translocation through the nuclear pore (27, 28). It is conceivable that the VirE2 protein has a role in the second step of import, the ss-T-DNA translocation, possibly by unfolding the long single-stranded DNA, to permit its passage through the pore. VirE2 has in fact been found to elongate the ss-DNA to which it is cooperatively bound by about 55% (12). In analogy to this proposed function, in ribonucleoprotein particles (RNP) the RNA is unfolded and elongated when passing through the nuclear pore (29). The transport of this complex is from the nucleus to the cytoplasm, but recent results revealed that nuclear export and nuclear import involve similar components (30). Another potential analogy between RNP export and T-complex import can be found in the directionality of the transport. The part of the RNP containing the 5' end of the RNA molecule is positioning the complex at the entrance of the central channel of the nuclear pore and passes through the pore first (31).

The C-terminal NLS of VirD2 may position the complex at the entrance of the nuclear pore and may pass through the pore first.

Integration of T-DNA into the plant genome. T-DNA is integrated randomly into the plant cell DNA (32, 33; for review see 34). Furthermore, the T-DNA has been described to integrate preferentially in transcriptionally active sequences (35, 36). Examination of the T-DNA-plant DNA junction sequences revealed that the right T-DNA border to which VirD2 is bound integrates more precisely than the left border suggesting that VirD2 might have a role in integration of the T-DNA (37, 38). Indeed, the VirD2 protein was shown recently to be involved in the ligation of the 5' end of the T-DNA into the plant cell DNA (39). While a particular mutation of *virD2* did not affect the efficiency of integration of the T-DNA, indicating that the ligation mediated by VirD2 is not the limiting step of T-DNA integration, it did affected the conservation of the 5' end of the T-DNA upon integration. Analysis of plants transformed with *A. tumefaciens* strains carrying this *virD2* mutation revealed that the integrated T-DNA had suffered truncations at the right border as compared to the original T-DNA (see Fig. 1). In contrast, transformation using the wild type strain resulted almost always in integration of the T-DNA up to the nucleotide which is linked to VirD2 (Fig. 1). These results allowed us to redesign a model of T-DNA integration according to which first the 3' end of the ss-T-DNA establishes the contact with the plant DNA and then the 5' end of the ss-T-DNA is ligated to the plant DNA (39). This illegitimate mode of recombination of the T-DNA with the plant DNA resembles the excision repair mechanism described for yeast (40) and humans (41). In these repair systems the 3' and 5' incisions are thought to be coordinated and therefore to result in deletions of short DNA fragments. These small deletions are the common features between T-DNA integration and excision repair (34).

The described ss-DNA binding properties of the VirE2 protein (8-13) may enable this protein to also play an important role in T-DNA integration; it may promote the establishment of the first contact between the 3' end of the T-DNA and the plant DNA. This was tested by determining the efficiency of T-DNA integration

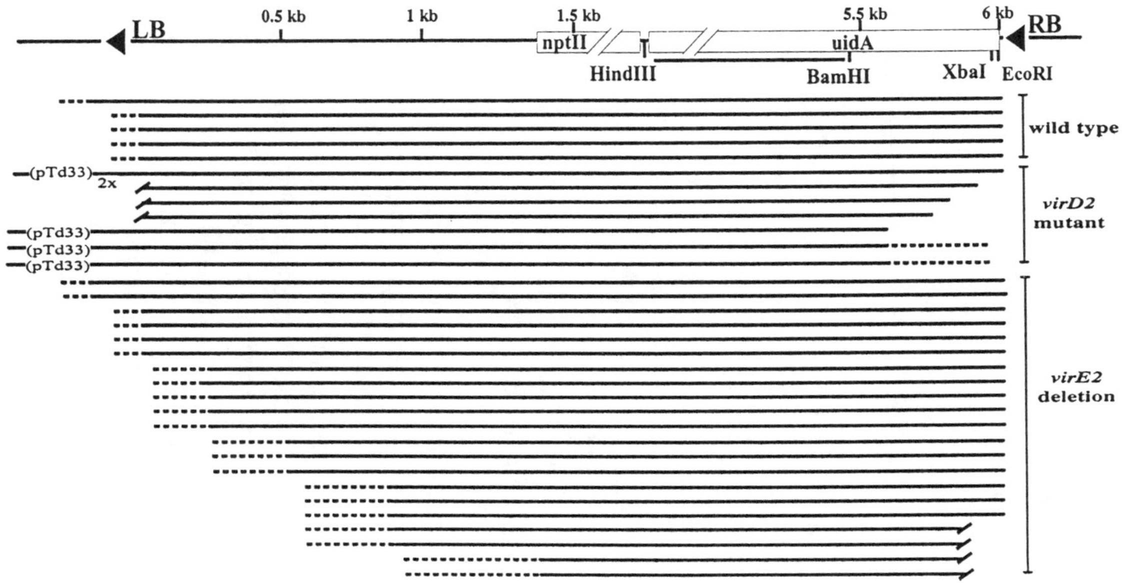

Figure 1. Representation of the pattern of T-DNA integrated into tobacco DNA after transformation with a wild type *A. tumefaciens* strain, a strain carrying a VirD2 protein with an arginine to glycine mutation at position 129 (39), and a strain carrying an almost complete deletion of the *virE2* gene (17). Inclined bar at the left end of three T-DNAs from a mutant virD2 strain: extent of T-DNA at the left junction was not analyzed. Inclined bars at the right end of four T-DNAs from a strain deleted for the *virE2* gene: the *Eco*RI site diagnostic for a complete right border-T-DNA junction is missing, but the precise junction site has not been determined.

following infection of plantlets with *A. tumefaciens* carrying a *virE2* deletion (the integration efficiency thereby is defined as ratio of T-DNA molecules which integrated into the genome out of the molecules transferred to the nucleus). The efficiency of integration of the *virE2* mutant was found to be comparable to the one of the wild type strain (17). The function of VirE2 thus does not seem to be important to promote integration of the T-DNA into the plant genome. The integration patterns of the T-DNA of plants transformed using the VirE2 minus strain were analyzed (17). Mainly severely truncated forms of the T-DNA were found integrated into the plant genome after T-DNA transfer mediated by the VirE2 deficient strain (Fig. 1) while the 3' end of the integrated part of the T-DNA transferred by the wild type strain contained short deletions of up to 50 nucleotides compared to the "original" T-DNA (Fig. 1). The most straightforward explanation for the function of VirE2, given its strong and cooperative binding to ss-DNA, is that it is coating the T-DNA on its way to the plant nucleus and that it thereby physically protects the T-DNA from plant nucleases.

An interesting question arises from this result; at which stage of the transfer is VirE2 coating the ss-T-DNA? One possibility is that a T-complex composed of the ss-T-DNA linked to VirD2 and coated by VirE2 forms in *A. tumefaciens* and then is transferred to the plant cell. This idea is contradicted by two findings. First, a strain mutated in the *virE2* gene can very efficiently transfer the T-DNA to transgenic plants constitutively expressing the VirE2 protein (18). Second, a *virE2* mutated strain can be efficiently complemented for T-DNA transfer by another *A. tumefaciens* strain producing the VirE2 protein (42, 9). Both T-DNA transfer and VirE2 transfer from these strains require the activity of the VirB proteins (43). Thus VirE2 protein molecules and the ss-T-DNA VirD2 complex may travel to the plant cell independently. This hypothesis is supported by the finding that export of VirE2, but not of the T-DNA, depends on VirE1 (44). Thus the mechanism of export of VirE2 and the T-DNA also has differences. The T-DNA/VirD2 complex and the VirE2 protein molecules may be transferred to the plant cell separately also during infection with wild type bacteria. VirE2 and the T-DNA would meet in the plant cell sufficiently early for the VirE2 to protect the T-DNA from plant nucleases. We found, in fact, that the pattern of T-DNA integration in plants transformed in coinoculation experiments of *A. tumefaciens* strains is corresponding to the pattern found for the transformation by the wild type strain (17). It is very difficult to define where exactly in the plant cell this complementation occurs. It can be speculated that VirE2 "waits" at the entry point of the T-DNA into the plant cell (wherever this would be) and helps the T-DNA to enter the plant cell or simply coats the ss-T-DNA entering the plant cell. This hypothesis is incompatible with the finding that VirE2 contains functional NLSs and therefore is localized predominantly in the nucleus. In fact if free VirE2 protein is located in the nucleus of the plant cell, it should coat the ss-T-DNA only there. It is at this point important to consider that the localization studies on VirE2 protein revealed that VirE2 localizes only partially into the plant cell nucleus; VirE2 protein was found also to be localized in the cytoplasm. Furthermore these studies have been done on protoplasts which do not have a cell wall. This part of the plant cell is necessary to establish the contact with the *A. tumefaciens* cell and may possibly be a region of the cell where VirE2 protein accumulates to exert its function for T-DNA import into the plant cell. It is therefore not yet possible to precisely determine the stage of the transfer at which the T-DNA is coated by VirE2.

Summary. *Agrobacterium tumefaciens* T-DNA transport to the plant has most probably evolved from bacterial conjugation. The plant cell represents a very different recipient cell as compared to the bacterial DNA acceptor. *A. tumefaciens* had to adapt to perform efficient T-DNA transfer to plants. Firstly, it had to develop strategies to recognize transformable plant cells. Secondly, it had to overcome the physical barrier represented by the plant cell wall. At the moment it is not known how *A. tumefaciens* succeeds to solve this latter problem and to transfer the T-DNA to the plant cytoplasm. In contrast, the strategies used by *A. tumefaciens* to perform the steps of the transfer which follow cell entrance of the T-DNA (nuclear targeting of the T-DNA and its integration into the plant genome) begin to be better understood. Critical roles for the successful T-DNA transfer are played by the VirD2 and VirE2 proteins which also travel from *A. tumefaciens* to the plant cell. The form of the T-DNA predominantly found in the plant cell is the single-stranded one: this form has been found associated with plant cells (7) and as major part of T-DNA molecules arriving in the nucleus (6). The function of VirE2 on the T-DNA is important for the conservation of T-DNA molecules in the plant cell and for the integration of full length T-DNA molecules into the plant genome (17). However, it is not known at which step of the transfer VirE2 associates with the T-DNA. Since it exerts its function exclusively inside the plant cell, it may be complexed to the T-DNA only in the plant cell. The VirD2 protein, the endonuclease which specifically cleaves the right and left border, is covalently bound to the 5' end of the T-DNA and travels with the T-DNA. The T-DNA is targeted to the nucleus most probably by the NLS localized in the C-terminal part of the VirD2 protein.

A. tumefaciens' strategy to get the T-DNA imported into the nucleus is to use the plant nuclear import machinery to import the VirD2 protein which then "carries" the T-DNA along. Plant enzymes, some of which are most probably involved in DNA repair, integrate the T-DNA into the plant genome. The VirE2 and VirD2 proteins do not seem to be directly involved in this integration step but are essential for the integration of full length T-DNA molecules. In summary, *A. tumefaciens* seems to have “learned” how to use the existing plant machinery for nuclear import, T-DNA integration and maybe also for import into the plant cell. To efficiently "conjugate" with the plant cell, *A. tumefaciens* also has “learned” to protect the T-DNA against the defense mechanism of the plant cell which acts against invading nucleic acids.

Acknowledgments

We enjoyed discussions with other members of the group. We acknowledge the receipt of not yet published information by A. Binns, S. Gelvin and W. Ream.

References

1. **Lessl, M. and E. Lanka.** 1994. Common mechanisms in bacterial conjugation and Ti-mediated T-DNA transfer to plant cells. Cell. **77**: 321-324.

2. **Greber, U. and H. Kasamatsu.** 1996. Nuclear transport of adenovirus and SV40. Trends in Cell Biol. **6**: 189-195.

3. **Stevenson, M.** 1996. Portals of entry: uncovering HIV nuclear transport pathways. Trends in Cell Biol. **6**: 9-15.

4. **Nigg, E.A., P.A. Baeuerle and R. Lührmann.** 1991. Nuclear Import-Export: In Search of Signals and Mechanisms. Cell **66**: 15-22.

5. **Noveiry, A.O., W.J. Lucas and R.L. Gilbertson.** 1994. Two proteins and plant DNA virus coordinate nuclear and plasmodesmatal transport. Cell **76**: 925-932.

6. **Tinland, B., B. Hohn and H. Puchta.** 1994. *Agrobacterium tumefaciens* transfers single-stranded T-DNA into the plant cell nucleus. Proc. Natl. Acad. Sci. USA **91**: 8000-8004.

7. **Yusibov, V.M., T.R. Steck, V. Gupta and S.B. Gelvin.** 1994. Association of single-stranded transferred DNA from *Agrobacterium tumefaciens* with tobacco cells. Proc. Natl. Acad. Sci. USA **91**: 2994-2998.

8. **Gietl, C., Z. Koukolíková-Nicola and B. Hohn.** 1987. Mobilization of T-DNA from *Agrobacterium* to plant cells involves a protein that binds single-stranded DNA. Proc.Natl. Acad. Sci. USA **84**: 9006-9010.

9. **Christie, P.J., J.E. Ward, S.C. Winans and E.W. Nester.** 1988. The *Agrobacterium tumefaciens virE2* gene product is a single-stranded DNA-binding protein that associates with T-DNA. J. Bacteriol. **170**: 2659-2667.

10. **Citovsky, V., G. de Vos and P. Zambryski.** 1988. Single-stranded DNA binding protein encoded by the *virE* locus of *Agrobacterium tumefaciens*. Science **240**: 501-504.

11. **Das, A.** 1988. *Agrobacterium tumefaciens virE* operon encodes a single-stranded DNA-binding protein. Proc. Natl. Acad. Sci. USA **85**: 2909-2913.

12. **Citovsky, V., M.L. Wong and P. Zambryski.** 1989. Cooperative interaction of *Agrobacterium* VirE2 protein with single-

stranded DNA: implications for the T-DNA transfer process. Proc. Natl. Acad. Sci. USA **86**: 1193-1197.

13. **Sen, P., G.J. Pazour, D. Anderson and A. Das.** 1989. Cooperative binding of *Agrobacterium tumefaciens* VirE2 protein to single-stranded DNA. J. Bacteriol. **171**: 2573-2580.

14. **Stachel, S.E., B. Timmerman and P. Zambryski.** 1986. Generation of single-stranded T-DNA molecules during the initial stages of T-DNA transfer from *Agrobacterium tumefaciens* to plant cells. Nature **322**: 706-712.

15. **Gardner, R.C. and V.C. Knauf.** 1986. Transfer of *Agrobacterium* DNA to plants requires a T-DNA border but not the *virE* locus. Science **231**: 725-727.

16. **Grimsley, N., B. Hohn, C. Ramos, C. Kado and P. Rogowsky.** 1989. DNA transfer from *Agrobacterium* to *Zea mays* or *Brassica* by agroinfection is dependent on bacterial virulence functions. Mol. Gen. Genet. **217**: 309-316.

17. **Rossi, L., B. Hohn and B. Tinland.** 1996. Integration of complete T-DNA units is dependent on the activity of VirE2 protein of *Agrobacterium tumefaciens*. Proc. Natl. Acad. Sci. USA. **93**: 126-130.

18. **Citovsky, V., J. Zupan J, D. Warnick and P. Zambryski.** 1992. Nuclear localization of *Agrobacterium* VirE2 protein in plant cells. Science **256**: 1802-1805.

19. **Herrera-Estrella, A., M. Van Montagu and K. Wang.** 1990. A bacterial peptide acting as a plant nuclear targeting signal: the amino-terminal portion of *Agrobacterium* VirD2 protein directs a beta-galactosidase fusion protein into tobacco nuclei. Proc. Natl. Acad. Sci. USA. **87**: 9534-9537.

20. **Tinland, B., Z. Koukolikova-Nicola, M.N. Hall and B. Hohn.** 1992. The T-DNA linked VirD2 contains two distinct functional nuclear localization signals. Proc. Natl. Acad. Sci. USA. **89**: 7442-7446.

21. **Citovsky, V., D. Warnick and P. Zambryski.** 1994. Nuclear import of *Agrobacterium* VirD2 and VirE2 proteins in maize and tobacco. Proc. Natl. Acad. Sci. USA. **91**: 3210-3214.

22. **Rossi, L., B. Hohn and B. Tinland.** 1993. VirD2 protein carries nuclear localization signals important for transfer of T-DNA to plants. Mol. Gen. Genet. **239**: 345-353.
23. **Shurvinton, C.E., L. Hodges and W. Ream.** 1992. A nuclear localisation signal in the *Agrobacterium tumefaciens* VirD2 endonuclease is important for tumor formation. Proc. Natl. Acad. Sci. USA. **89**: 11837-11841.
24. **Koukolíková-Nicola, Z., D. Raineri, K. Stephens, C. Ramos, B. Tinland, E.W. Nester and B. Hohn.** 1993. Genetic analysis of the *vir*D operon of *Agrobacterium tumefaciens*: a search for functions involved in the transport of T-DNA into the plant cell nucleus and in T-DNA integration. J. Bacteriol. **175**: 723-731.
25. **Burkrinsky, M., S. Haggerty, M. Dempsey, N. Sharova, A. Adzhubel, L. Spitz, P. Lewis, D. Goldfarb, M. Emermen and M. Stevenson.** 1993. A nuclear localization signal within HIV-1 matrix protein that governs infection of non-dividing cells. Nature **365**: 666-668.
26. **LaCasse, E.C. and Y.A. Lefebvre.** 1995. Nuclear localization signals overlap DNA or RNA-binding domains in nucleic acid-binding proteins. Nucleic Acids Res. **23**: 1647-1656.
27. **Newmeyer, D.D. and D.J. Forbes.** 1988. Nuclear import can be separated into distinct steps in vitro: nuclear pore binding and translocation. Cell. **52**: 641-653.
28. **Richardson, W.D., A.D, Mills, S.M. Dilworth, R.A. Laskey and C. Dingwall.** 1988. Nuclear protein migration involves two steps: rapid binding at the nuclear envelope followed by slower translocation through the nuclear pores. Cell. **52**: 655-664.
29. **Mehlin, H., B. Daneholt and U. Skoglund.** 1992. Translocation of a specific premessenger ribonucleoprotein particle through the nuclear pore studied with electron microscope tomography. Cell. **69**: 605-613.
30. **Gerace, L.** 1995. Nuclear export signals and the fast track to the cytoplasm. Cell **82**:341-344.
31. **Mehlin, H., B. Daneholt and U. Skoglund.** 1995. Structural interaction between the Nuclear Pore Complex and a specific translocating RNP Particle. J. Cell Biol. **129**: 1205-1216.

32. **Chyi, Y-S., R.A. Jorgensen, D. Goldstein, S.D. Tanksley and F.L. Loaiza-Figueroa.** 1986. Locations and stability of *Agrobacterium*-mediated T-DNA insertions in the *Lycopersicon* genome. Mol. Gen. Genet. **204**: 64-69.

33. **Wallroth, M., A.G.M. Gerats, S.G. Rogers, R.T. Fraley and R.B. Horsch**. 1986. Chromosomal localization of foreign genes in Petunia hybrida. Mol Gen Genet. **202**: 6-15.

34. **Tinland, B. and B. Hohn.** 1995. Recombination between prokaryotic and eukaryotic DNA: integration of *Agrobacterium tumefaciens* T-DNA into the plant genome, p. 209-229. *In* J. K. Setlow (ed.) Genetic Engineering, Principles and Methods. Plenum New York.

35. **Koncz, C., N. Martini, R. Mayerhofer, Z. Koncz-Kalman, H. Körber, G.P. Redei and J. Schell.** 1989. High-frequency T-DNA-mediated gene tagging in plants. Proc. Natl. Acad. Sci. USA. **86**: 8467-8471.

36. **Herman, L., A. Jacobs, M. Van Montagu and A. Depicker.** 1990. Plant chromosome/marker gene fusion assay for study or normal and truncated T-DNA integration events. Mol. Gen. Genet. **224**: 248-256.

37. **Gheysen, G., R. Villarroel and M. van Montagu.** 1991. Illigitimate recombination in plants: a model for T-DNA integration. Genes & Dev. **5**: 287-297.

38. **Mayerhofer, R., Z. Koncz-Kalman, C. Nawrath, G. Bakkeren, A. Crameri, A. Angelis, G.P. Redei, J. Schell, B. Hohn and C. Koncz.** 1991. T-DNA integration: a mode of illegitimate recombination in plants. EMBO J. **10**: 697-704.

39. **Tinland, B., F. Schoumacher, V. Gloeckler, A.M. Bravo Angel and B. Hohn.** 1995. The *Agrobacterium tumefaciens* Virulence D2 protein is responsible for precise integration of T-DNA into the plant genome. EMBO J. **13**: 5764-5771.

40. **Prakash, S., P. Sung and L. Prakash.** 1993. DNA repair genes and proteins of *Saccharomyces cerevisiae*. Annu. Rev. Genet. **27**: 33-70.

41. **Hoeijmakers, J.H.J. and D. Bootsma.** 1994. Incisions for excision. Nature. **371**: 654-655.

42. **Otten, L., H. De Greve, J. Leemans, R. Hain, P.J.J. Hooykaas and J. Schell.** 1984. Restoration of virulence of *vir* region mutants of *Agrobacterium tumefaciens* strain B6S3 by coinfection with normal and mutant *Agrobacterium* strains. Mol. Gen. Genet. **195** : 159-163.

43. **Binns, A., C. Beaupré and E.M. Dale.** 1995. Inhibition of VirB-mediated transfer of diverse substrates from *Agrobacterium tumefaciens* by the IncO plasmid RSF1010. J. Bacteriol. **177**: 4890-4899.

44. **Sundberg, C., L. Meek, K. Carroll, A. Das and W. Ream.** 1996. VirE1 protein mediates export of the single-strand DNA-binding protein VirE2 from *Agrobacterium tumefaciens* into plant cells. J. Bacteriol. **178**:1207-1212.

CHAPTER 4

Ecological Significance of Avirulence in *Agrobacterium*

Patrice Dion, Christian Bélanger* and Claude Marquis

*Institut Maurice Lamontagne, Fisheries and Oceans Canada, Mont-Joli, Québec, Canada, G5H 3Z4
and
Recherches en sciences de la vie et de la santé, Université Laval, Québec, Québec, Canada G1K 7P4

Clones constituting the natural populations of *Agrobacterium* can be distinguished from each other on the basis of chromosome- or plasmid-encoded characters (15). Distinction between species of the genus *Agrobacterium* is achieved mainly on the basis of chromosomal markers, which are considered to be more stable than traits determined by agrobacterial Ti or Ri plasmids. Indeed, variability in phenotypic traits encoded by agrobacterial plasmids is readily demonstrated in the laboratory, as certain strains can be observed to lose or gain plasmids, with a consequent loss or acquisition of virulence. Whereas plasmid loss and transfer are likely to occur in nature also, these phenomena cannot account, by themselves, for all of the variability which is encountered in natural populations of *Agrobacterium*.

Opines are thought to play a central role in the relationship between virulent agrobacteria and their transformed plant host, since these derivatives of amino acids or sugars act as chemical mediators of parasitism. Opines are synthesized by the transformed plant cell, at the direction of genes contained in a fragment of the Ti or Ri plasmid which is transferred from the bacterium to the plant. They diffuse in and around the tumor, where they are recognized as substrates by bacteria harboring plant-transforming plasmids (8). Genes located in non-transferred regions of these plasmids code for specific catabolic systems responsible for the uptake and degradation of opines (18, 21). Generally, virulent agrobacteria utilize those types of opines which are produced by the tumors they incite. In addition to virulent agrobacteria carrying Ti or Ri plasmids, a wide variety of soil and plant-associated microorganisms are capable of utilizing opines (Table 1). These include fungi of the genera *Fusarium* and *Cylindrocarpon* (1), Gram-positive bacteria belonging to the genus *Arthrobacter* (14,

Table 1. Properties of opine-utilizing microorganisms.

Type of microorganism	Origin	Type of opines utilized[a]		
		Iminodiacids[b]	Mannityl opines	Phosphorylated sugars
Agrobacterum spp.[c]	Plants[d], soil	+	+	+
Rhizobium meliloti	Plants	+	NS	NS
Pseudomonas spp.[e]	Plants, soil	+	+	NS
Arthrobacter spp.	Soil	+	+	NS
Pseudomonas spp.[f]	Marine molluscs	+	NS	NS
Cylindrocarpon spp., *Fusarium* spp.[g]	Plants	+	+	NS

[a] +, utilization of opines of the corresponding type; NS, utilization of opines not shown.
[b] Includes octopine and nopaline.
[c] Includes virulent and avirulent strains.
[d] Includes crown gall tumors and healthy plant tissue.
[e] Includes fluorescent and nonfluorescent strains.
[f] Marine bacteria requiring sodium chloride for growth.
[g] Saprophytic fungi in crown gall tumors.

17), fluorescent and nonfluorescent pseudomonads (5, 13, 14), rhizobia (4) and avirulent agrobacteria (5, 13). In some of these other types of opine utilizers, such as *Pseudomonas* for example, the genes of opine catabolism differ, with respect to both structure and regulation, from their Ti plasmid-encoded counterparts (3, 14). In others, such as *Rhizobium meliloti* which is taxonomically very close to *Agrobacterium*, the genes for catabolism of one of the opines, called octopine, are regulated similarly and strongly homologous to the corresponding catabolic genes present on the Ti plasmid. As in *A. tumefaciens*, the *R. meliloti* genes of octopine catabolism are plasmid-encoded (4).

Conceivably, two distinct mechanisms could contribute to the emergence of opine-utilizing, avirulent agrobacteria. Under laboratory conditions, spontaneous mutations resulting in the acquisition of capacity to catabolize opines are known to occur in agrobacteria devoid of a Ti plasmid (12). Alternatively, derivatives of the Ti plasmid could arise in which genetic elements determining virulence, such as the *vir* or T regions, would have been lost or disarmed. Such derivatives, which would still harbor regions specifying opine catabolism, could be produced by a variety of mechanisms such as insertion of foreign DNA into genes essential for virulence, point mutation or deletion of such genes, or alternatively cointegration of the Ti plasmid with another indigenous replicon followed by molecular rearrangements.

Several independent observations suggest that, indeed, such disarmed derivatives of the Ti plasmids are produced in nature. One of these, alluded to above, is the discovery of genes of octopine catabolism in a megaplasmid of *R. meliloti*. The Ti plasmid has the capacity to be transferred to *R. meliloti* and to replicate in this new host. It would then have the opportunity to recombine with resident plasmids, with the consequent loss of certain functions encoded by the original Ti plasmid. Along the same lines, it has also been noted that one of the two plasmids from the biological control agent *Agrobacterium radiobacter* K84 carried genes for catabolism of another opine, nopaline. This plasmid lacks both the *vir* and T-DNA regions but may share a common ancestry with the Ti plasmid from the nopaline-type strain C58 (6, 9).

Sensitivity to acetosyringone in nopaline-type strains. Initially, it seemed possible that induction of the *vir* genes might trigger rearrangements of the Ti plasmid leading to loss of oncogenicity. VirD1 and VirD2 possess endonuclease activity responsible for the excision of the T region from the Ti plasmid and it was hypothesized that deleted derivatives of the Ti plasmid, lacking a T region, might arise as a result of this endonucleolytic cleavage. To test this hypothesis, it was decided to expose *A. tumefaciens* C58 to *vir* gene inducing conditions, and then to examine individual clones from this treated culture for tumor inducing capacity.

As compared to those from a control culture, bacteria that were exposed to 60 μM of the *vir* gene inducer acetosyringone exhibited a growth retardation response, characterized by a prolonged lag phase preceding initiation of exponential growth. Up to 100% of those clones that eventually arose in the acetosyringone-treated cultures had lost their ability to induce tumors on tomato and other host plants. These avirulent mutants were nonresponsive to acetosyringone, in the sense that they did not exhibit the *vir* gene induction nor the growth retardation responses (see above) to this phenolic compound. However, most of these mutants retained the capacity to utilize nopaline.

The two phenomena characteristic of acetosyringone-treated cultures, growth retardation and production of avirulent clones, appeared to be related. This was suggested, in particular, by the fact that no avirulent mutants were detected during the prolonged lag phase observed in the presence of acetosyringone, whereas, as mentioned above, most of the clones recovered at the end of exponential growth were avirulent.

The negative effects on growth and oncogenic potential were dependent on *vir* gene induction. In the presence of acetosyringone, these effects were not observed at pH 7, a pH which is not conducive to induction of the *vir* genes. Furthermore, a correlation was observed between the activity of analogs of acetosyringone as *vir* gene inducers and as effectors of both the growth retardation response and avirulent mutant production. Sinapinic acid, which was a good *vir* gene inducer, had effects on strain C58 cultures which were similar to those of acetosyringone. On the other hand, syringaldehyde promoted only a weak *vir* gene induction and accordingly did not cause marked retardation of growth or accumulation of avirulent mutants (10).

The effects of acetosyringone on growth and oncogenic potential were not observed solely for strain C58, but were noted also for a different nopaline-type strain, T37. In contrast with this, the octopine-type strain B6S3 and strain C58 derivatives harboring, in place of the native plasmid pTiC58, the octopine-type plasmid pTiA6NC or the mannopine-type plasmid pTiBo542 did not exhibit growth inhibition or accumulation of avirulent mutants in the presence of acetosyringone. Acetosyringone also did not affect growth or oncogenic potential of *Agrobacterium rhizogenes* A4 (10; I. Loubens and P. Dion, unpublished).

The altered characteristics of the acetosyringone-resistant clones of strain C58 were related to modifications of the Ti plasmid. This was demonstrated by transfer of the Ti plasmid from these clones to a Ti plasmidless recipient, and subsequent examination of the phenotype of the resulting transconjugants. As did the Ti plasmid donor clones, these transconjugants failed to exhibit the growth retardation and *vir* gene induction responses to acetosyringone. Furthermore, virulence of many of the acetosyringone-resistant clones was restored by complementation with clones containing intact *virA* or *virG*. Examination of the altered Ti plasmids from the acetosyringone-resistant clones showed that some of these plasmids had undergone deletions which encompassed most of the *vir* region and, in some cases, extended into the T-DNA. In other cases, the insertion sequence element IS*426* was detected in *virA* or *virG* of the defective Ti plasmids (Fig. 1). These two genes determine the synthesis of the two-component regulatory system responding to acetosyringone and other *vir* gene inducers. Thus, the two molecular events responsible for inactivation of the Ti plasmid in the acetosyringone-resistant clones, deletion and DNA insertion, differed with respect to their magnitude and specificity, but both provoked inactivation of the *virA-virG* regulatory function (11).

These studies on the effect of acetosyringone on the genetic stability of *A. tumefaciens* strains were initiated on the premise that specific endonucleolytic activity, expressed under *vir* gene inducing

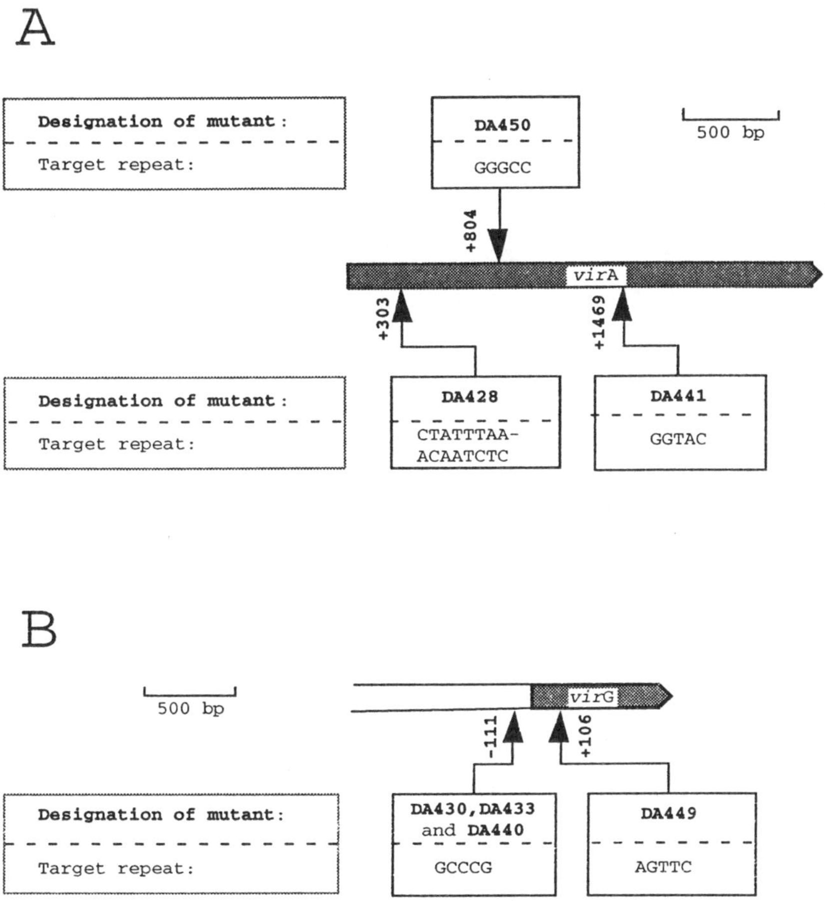

Figure 1. Localization of sites of integration of IS426 in *virA* (A) or *virG* (B) of avirulent mutants recovered from a culture of *A. tumefaciens* C58 treated with 60 μM acetosyringone. The target *vir* gene DNA is represented by an horizontal bar, and the direction of transcription of this target gene is given. For each individual mutant, an arrow indicates the position of IS426 with respect to the translation initiation site. Arrows are placed above or below the bar representing target DNA, depending on orientation of the element. In a box connected to each of the arrows, the designation of the corresponding mutant and the DNA repeat identified at the target site are indicated.

conditions, would provoke excision and subsequent loss of the T region of the Ti plasmid (see above). This particular hypothesis proved untrue, since the targets of mutagenic activity revealed in the presence of acetosyringone were identified as *virA* and *virG*, which are the mediators of the *vir* gene induction response. However, the results obtained in the course of these experiments suggested an alternative mechanism by which avirulent, but opine-utilizing, agrobacteria could be generated in the course of infection of the host plant.

Role of endogenous insertion elements in mutagenesis. A striking conclusion from the analysis of acetosyringone-resistant mutants of strain C58 was that, in most of the cases examined, mutagenesis appeared to be due to the activity of the endogenous element IS*426*. This 1.3-kilobase element exists in three copies in the chromosome of the wild-type C58 strain (19). It shares homology with IS*Rm1* (20), an element which transposes preferentially into the nitrogen fixation genes of *Rhizobium meliloti* (16). It is normally absent from the Ti plasmid of strain C58 but has been detected in the Ti plasmids of certain mutants where it was found, in particular, in the agrocin sensitivity (7), nopaline catabolism (11) and T (19) regions. We have observed that both the wild-type strain C58, which does not harbor IS426 in the Ti plasmid, and the derivative C58F, which contains an additional copy of the element in the Ti plasmid region of nopaline catabolism, produced avirulent mutants at an equally high frequency in the presence of acetosyringone. Thus, prior presence of the IS element in the Ti plasmid does not seem to be a critical factor in determining the frequency of insertions in *virA* or *virG*.

As shown above, activity of IS*426* appeared to be involved in the generation of most acetosyringone-resistant mutants of strain C58 that were analyzed. As had been the case for the C58 mutants, the T37 mutants that were recovered following growth in the presence of acetosyringone were defective in *virA* or *virG*. Further analysis of some of these T37 mutants suggested that, in the case of this strain also, mutations to acetosyringone resistance and avirulence resulted from insertion of an IS element in *virA* or *virG*. Strain T37 showed only little homology to an IS*426* probe, suggesting that, in this strain, the IS element that might be involved in mutagenesis differed, to some extent at least, from IS*426* (11).

Two distinct phenomena, mutagenesis and mutant selection, participated in the generation of up to 99% avirulent clones in cultures of strain C58 treated with acetosyringone. Selection of acetosyringone-insensitive mutants was facilitated by inhibition of growth of the wild-type cells in the presence of the *vir* gene inducer (see above). Whether or not *vir* gene induction also enhanced mutagenesis *per se* remains currently unclear. The possibility that acetosyringone could have a general effect on transposition activity of IS *426* was examined by incubating strain C58 in induction medium containing acetosyringone and 2 g per liter of yeast extract. It was thought that perhaps stimulation of mutagenesis under *vir* gene inducing conditions would lead to the accumulation of auxotrophic mutants. However, no auxotrophic mutants were detected in the culture at the end of the incubation period, suggesting that, in fact, exposure to acetosyringone does not lead to a general increase in the frequency of mutagenesis (C. Marquis and P. Dion, unpublished). The possibility remains that acetosyringone has a more specific effect on mutant production, so that insertion of IS*426* in certain regions of the Ti plasmid could be more frequent in the presence of the *vir* gene inducer than in its absence.

Accumulation of avirulent mutants as a consequence of crown-gall tumor induction. Induction media containing acetosyringone or other *vir* gene inducers are designed to mimic conditions prevailing at the wound site and it is generally believed that bacterial behavior, as it is observed in such laboratory media, reflects the events occurring during plant infection. Whether or not this general rule applied to the particular case of appearance of the acetosyringone-resistant mutants was examined by infecting axenic apple plants grown in tissue culture with various strains of *Agrobacterium*. In these experiments apple was chosen, first, as a source of test bacteria and, in a second step, as a host plant in which the effect of tumorigenesis on the tumor-inciting bacterium would be examined (Fig. 2A). Indeed, it is notoriously difficult to recover virulent agrobacteria from apple tumors, and the possibility of a marked genetic instability of the inciting bacterium, as a direct consequence of its interaction with the apple host, was considered as a possible explanation. Some other natural hosts of *Agrobacterium*, namely blackberry, pear and cherry, were also included in the study.

Five apple isolates out of nine tested yielded avirulent clones following their reintroduction into apple plants grown in tissue culture (Fig. 2). Two of these five unstable strains, D10B/87 and F27/93,

A

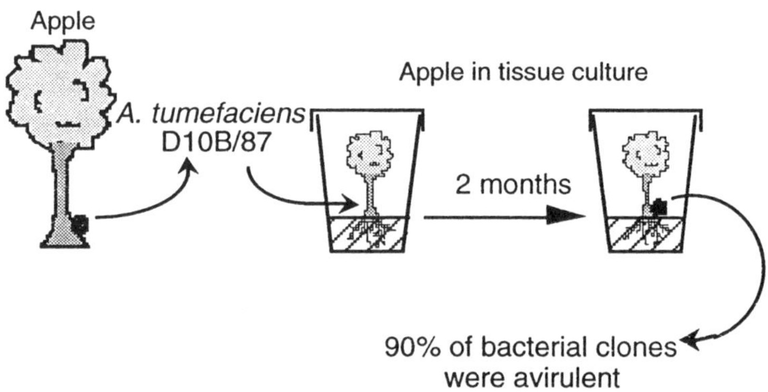

B

Other unstable strains	% of avirulent mutants recovered
F27/93	69
S120/93	69
S12/93	21
J5/75	7

Figure 2. Recovery of avirulent mutants of *A. tumefaciens* from apple tumors. (A) Diagrammatic representation of the experiment. *A. tumefaciens* D10B/87 is a natural isolate isolated from an apple tree grown in an orchard. This isolate was then used to infect apple trees grown in tissue culture. Bacteria were reisolated from the resulting experimental tumors and tested for virulence on tomato plants. (B) Results obtained when a procedure similar to that shown in (A) was applied to other apple isolates.

also produced avirulent clones during infection of pear and one, strain F27/93, showed some instability on blackberry. Depending on the strain, avirulent clones represented between 7 and 90% of the total number of clones recovered two months after inoculation of apple plants. Two cherry isolates, two blackberry isolates and the acetosyringone-sensitive strain C58 (see above) did not show signs of genetic instability following their inoculation to apple plants (2).

As had been the case for the acetosyringone-resistant mutants of strain C58 (see above), avirulent clones of strain D10B/87 were altered in the Ti plasmid. In separate clones obtained from the same tumor, the mutation responsible for avirulence was either identical or else very similar in nature. On the other hand, distinct mutations were found when clones obtained from different tumors were compared. Examination of a first set of 15 avirulent clones obtained from one particular tumor showed that all of these clones carried a point mutation in *virG*. For two other sets of avirulent clones, recovered from different tumors, the mutation identified corresponded to a deletion of the complete *vir* region and a point mutation in the *virG* promoter region, respectively. Upon incubation in minimal medium containing acetosyringone, strain D10B/87 also yielded avirulent mutants, but in this case a variety of mutants, carrying insertions in *virA* or point mutations in *virG*, were obtained from the same culture (2).

In the case of strain C58, growth inhibition accompanied by selection was found to be an essential component of the process leading to accumulation of avirulent mutants in acetosyringone-treated cultures (see above). The fact that strain D10B/87 also accumulated avirulent mutants in the presence of acetosyringone suggested that, like strain C58, this apple isolate was acetosyringone-sensitive, allowing for the previously identified mechanism of avirulent mutant selection to occur in a synthetic medium containing the *vir* gene inducer. By extrapolation, the hypothesis was formulated that selection played a role in the accumulation of avirulent mutants in infected apple plants. To test this possibility, apple plants were coinfected with a mixture of the wild-type strain D10B/87 and an avirulent mutant carrying a point mutation in *virG*. It was found that within two to five weeks, depending on the initial proportion of wild-type and mutant cells, the *virG* mutant outcompeted the wild-type strain. These results suggested that, indeed, a *virG* mutant of strain D10B/87 possessed a better ability to be maintained or replicate in the infected plant, as compared to the wild-type strain. Thus, as was the case for acetosyringone-treated *in vitro* cultures of *A. tumefaciens*, selection appears to be a component of the process leading to accumulation of avirulent mutants in infected stems.

In the case of *in vitro* exposure of strain D10B/87 to *vir* gene inducers, the outcome of the selection process is the generation of a wide variety of mutants from one particular culture. On the other hand, most mutants arising from a given apple tumor are identical, suggesting that they are of clonal origin (see above). Presumably, and considering that mutant selection takes place in the course of incubation, this difference in the extent of diversity of mutants produced *in vitro* and *in planta* reflects an equivalent difference in the number of independent mutants present at an early stage of incubation in medium with acetosyringone or in stems. To account for a greater number of different mutant types early during *in vitro*, as opposed to *in planta*, incubation, one might suggest that cells of strain D10B/87 are genetically more unstable when they grow in minimal medium with acetosyringone than when they are present in apple stems.

The purpose of the experiments conducted *in planta* was to determine whether or not avirulent mutants were produced under conditions that mimicked, as closely as possible, conditions occurring during natural agrobacterial infection. As described above, the answer to this question was yes, although at least two limitations were apparent. In particular, more mutants were obtained on apple than on the two other host plants examined, pear and blackberry. A second restriction was that the acetosyringone-sensitive strain C58 did not produce avirulent mutants upon its inoculation to apple. In fact, the only bacteria showing instability on apple had originally been recovered from apple tumors (see above).

The observation that agrobacterial strains originally isolated from apple exhibited massive production of avirulent mutants upon their experimental reinoculation to apple grown in tissue culture might appear contradictory. It should be noted, however, that the tissue culture conditions might amplify the genetic instability exhibited by agrobacteria inoculated to apple plants. While strain D10B/87 produced up to 90% avirulent mutants following its inoculation to apple plants prepared in tissue culture (see above), no mutants were recovered from tumors induced on apple plants maintained in the greenhouse. One possible explanation for this striking difference is that plants grown in tissue culture might make more nutrients available for agrobacterial growth, in the form of sugars and mineral salts taken from the culture medium, than plants grown in the greenhouse. Thus, the first type of plants might sustain better bacterial multiplication than the second, and as such provide conditions more favorable to the selection of avirulent mutants. A second factor that might perhaps contribute to insure a certain degree of natural stability to the Ti plasmids is the ability of these plasmids to be transferred between bacteria by conjugation. It might be envisioned that a particular Ti plasmid might confer a high degree of sensitivity to acetosyringone when in combination with certain chromosomal backgrounds, but much less sensitivity when associated with other backgrounds. In this sense, certain agrobacterial strains could possibly act as natural reservoirs of intact Ti plasmids.

Summary. The results summarized in this chapter reveal the existence of factors, associated with natural infection processes, that favor the expression of genetic instability in *A. tumefaciens*. We hypothesize that these factors contribute to reduce the ratio of virulent to avirulent strains in nature. They favor the accumulation of bacteria carrying modified Ti plasmids which exhibit alterations in certain essential virulence genes. The specific nature of these alterations, whether they are deletion, insertion or point mutations, may depend on the incubation conditions and mechanisms of gene instability in any particular strain. For example, in strains, such as strain C58, which harbor very active IS elements, insertional mutants are likely to be produced and selected. On the other hand, in strains that would be devoid of such active IS elements *vir* gene inactivation by point mutation might be a favored mechanism.

Following the initial mutagenesis event occurring in the *vir* region, further evolution of the inactivated Ti plasmid would then generate agrobacterial megaplasmids which would appear only distantly related to the original Ti plasmid conferring virulence. Genes of opine utilization carried by these megaplasmids might allow their host bacteria to efficiently compete for opines produced in crown-gall tumors.

Acknowledgments

This work was supported by grants from the Natural Sciences and Engineering Research Council of Canada and the Québec Fonds pour la formation de chercheurs et l'aide à la recherche.

References

1. **Beauchamp, C. J., W. S. Chilton, P. Dion, and H. Antoun.** 1990. Fungal catabolism of crown gall opines. Appl. Environ. Microbiol. **56**: 150-155.
2. **Bélanger, C., M. L. Canfield, L. W. Moore, and P. Dion.** 1995. Genetic analysis of nonpathogenic *Agrobacterium tumefaciens* mutants arising in crown gall tumors. J. Bacteriol. **177**:3752-3757.
3. **Bergeron, J., R. A. MacLeod, and P. Dion.** 1990. Specificity of octopine uptake by *Rhizobium* and *Pseudomonas* strains. Appl. Environ. Microbiol. **56**: 1453-1458.
4. **Bergeron, J., C. Beaulieu, R. C. Levesque, A. Kondorosi, and P. Dion.** 1993. Homology between the genes of octopine catabolism in *Rhizobium meliloti* A3 and corresponding genes on the Ti plasmid. Can. J. Microbiol. **39**: 1041-1050.
5. **Canfield, M. L., and L. W. Moore.** 1991. Isolation and characterization of opine-utilizing strains of *Agrobacterium tumefaciens* and fluorescent strains of *Pseudomonas* spp. from rootstocks of *Malus*. Phytopathology, **81**: 440-443.
6. **Clare, B. G., A. Kerr, and D. A. Jones.** 1990. Characteristics of the nopaline catabolic plasmids in *Agrobacterium* strains K84 and K1026 used for biological control of crown gall disease. Plasmid **23**: 126-137.

7. **Cooksey, D. A. 1986.** A spontaneous insertion in the agrocin sensitivity region of the Ti-plasmid of *Agrobacterium tumefaciens* C58. Plasmid **16**: 222-224.

8. **Dessaux, Y., A. Petit, and J. Tempé.** 1992. Opines in *Agrobacterium* biology, p. 109-136. *In* D. P. S. Verma (ed.), Molecular signals in plant-microbe communications. CRC Press, Inc., Boca Raton, Fla.

9. **Farrand, S. K., C.-L. Wang, S.-B. Hong, S. B. O'Morchoe, and J. E. Slota.** 1992. Deletion derivatives of pAgK84 and their use in the analysis of *Agrobacterium* plasmid functions. Plasmid **28**: 201-212.

10. **Fortin, C., E. W. Nester, and P. Dion.** 1992. Growth inhibition and loss of virulence in cultures of *Agrobacterium tumefaciens* treated with acetosyringone. J. Bacteriol. **174**: 5676-5685.

11. **Fortin, C., C. Marquis, E. W. Nester, and P. Dion.** 1993. Dynamic structure of *Agrobacterium tumefaciens* Ti plasmids. J. Bacteriol. **175**: 4790-4799.

12. **LaPointe, G., C. S. Nautiyal, W. S. Chilton, S. K. Farrand, and P. Dion.** 1992. Spontaneous mutation conferring the ability to catabolize mannopine in *Agrobacterium tumefaciens*. J. Bacteriol. **174**: 2631-2639.

13. **Nautiyal, C. S., and P. Dion.** 1990. Characterization of the opine-utilizing microflora associated with samples of soil and plants. Appl. Environ. Microbiol. **56**: 2576-2579.

14. **Nautiyal, C. S., P. Dion, and W. S. Chilton.** 1991. Mannopine and mannopinic acid as substrates for *Arthrobacter* sp. strain MBA209 and *Pseudomonas putida* NA513. J. Bacteriol. **173**: 2833-2841.

15. **Nesme, X., C. Ponsonnet, C. Picard, and P. Normand.** 1992. Chromosomal and pTi genotypes of agrobacterium strains isolated from *Populus* tumors in two nurseries. FEMS Microbiol. Ecol. **101**: 189-196.

16. **Ruvkun, G. B., S. R. Long, H. M. Meade, R. C. van den Bos, and F. M. Ausubel.** 1982. ISRm1: a *Rhizobium meliloti* insertion sequence that transposes preferentially into nitrogen fixation genes. J. Mol. Appl. Genet. **1**: 405-418.

17. **Tremblay, G., R. Gagliardo, W. S. Chilton, and P. Dion.** 1987. Diversity among opine-utilizing bacteria: identification of coryneform isolates. Appl. Environ. Microbiol. **53**: 1519-1524.

18. **Valdivia, R. H., L. Wang, and S. C. Winans.** 1991. Characterization of a putative periplasmic transport system for octopine accumulation encoded by *Agrobacterium tumefaciens* Ti plasmid pTiA6. J. Bacteriol. **173**: 6398-6405.

19. **Vanderleyden, J., J. Desair, C. De Meirsman, K. Michiels, A. P. Van Gool, M. D. Chilton, and G. C. Jen.** 1986. Nucleotide sequence of an insertion sequence (IS) element identified in the T-DNA region of a spontaneous variant of the Ti-plasmid pTiT37. Nucleic Acids Research **14**: 6699-6709.

20. **Watson, R. J., and R. Wheatcroft.** 1991. Nucleotide sequence of *Rhizobium meliloti* insertion sequence IS*Rm*1: homology to IS2 from *Escherichia coli* and IS426 from *Agrobacterium tumefaciens*. DNA Sequence-J. DNA Sequencing and Mapping, **2**: 163-172.

21. **Zanker, H., G. Lurz, U. Langridge, P. Langridge, D. Kreusch, and J. Schröder.** (1994) Octopine and nopaline oxidases from Ti plasmids of *Agrobacterium tumefaciens*: molecular analysis, relationship, and functional characterization. J. Bacteriol. **176**: 4511-4517.

CHAPTER 5

Transcriptional Regulation of Conjugal Transfer Genes of Octopine-type Ti Plasmids

Stephen C. Winans, Lu Wang, Paulette Dwen, Clay Fuqua, Margret I. Moré, Juliane Alt-Mörbe, Joel Styker, and Michelle Burbea

Section of Microbiology
Cornell University
Ithaca, New York 14853

Ti plasmids of *Agrobacterium tumefaciens* have two independent DNA transfer systems. As described elsewhere in this volume, these bacteria can transfer discrete fragments of oncogeneic DNA (T-DNA) into the nuclei of infected plant cells; this process requires the products of approximately 25 *vir* genes (36). In addition, Ti plasmids are capable of autonomous conjugal transfer between Agrobacteria, and the products of approximately 16 *tra* genes are required for this process (1). *vir* genes and *tra* genes, although distantly related evolutionarily, are generally considered to function independently (2, 6, 9, see however, ref. 15, 28).

The *tra* regulon of octopine-type Ti plasmids is subject to rather complicated transcriptional regulation. It is positively regulated in a two-step transcriptional cascade involving three Ti plasmid-encoded proteins: OccR, TraR and TraI (10). OccR is stimulated by the tumor-released compound octopine, and causes conjugation to occur only in the presence of this compound (17). TraR and TraI are homologous to LuxR and LuxI, respectively, of *Vibrio fischeri* and, as predicted, cause conjugation to occur only at high cell densities (14). Ti plasmid conjugation therefore, occurs only in, on, or near densely populated crown gall tumors. It is far from clear why controlling conjugation in this fashion would be adaptive.

OccR is a LysR-type transcriptional regulator (17). In the presence of octopine, OccR positively regulates the 14 gene *occ* operon (see below). OccR is also a repressor of the *occR* gene, which

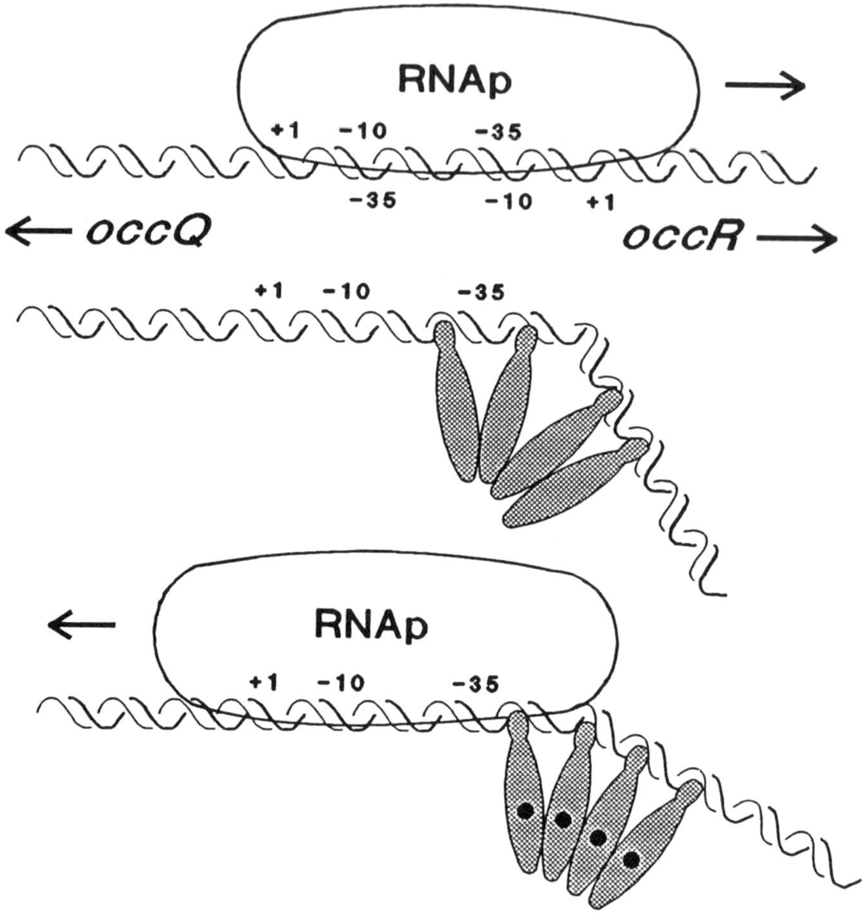

Figure 1. Ligand-responsive DNA bending by OccR protein (shaded). In the absence of OccR, the *occR* promoter is active, while the *occQ* promoter is silent. OccR protein represses *occR* but does not activate *occQ*. Under these conditions, OccR protects five helical turns and causes a high angle DNA bend. Octopine (black dots) causes OccR to protect only four helical turns, and relaxes the angle of the DNA bend. Octopine also converts OccR from an inactive form to a form able to activate *occQ* transcription (Reprinted, by permission, from ref. 35).

is transcribed divergently from the *occ* operon (17). Importantly, repression of *occR* occurs both in the presence and absence of octopine, suggesting that binding of OccR to its operator is not affected by this ligand.

The fact that octopine is essential for *occ* activation indicates that octopine must alter the conformation of bound OccR protein. This prediction was borne out in a series of *in vitro* reconstitution experiments (33). OccR protein protects a rather long region of DNA that overlapped the *occR* promoter and lies directly upstream of the *occ* promoter. OccR bound this site both in the presence and absence of octopine, and the affinity of the protein for this site was weakened by octopine only about two fold. However, octopine did alter the properties of bound OccR in several interesting ways. First, octopine converted OccR from in an inactive form to a form able to activate the *occQ* promoter. Second, octopine shortened the OccR footprint from five turns of the helix to four turns. The change in footprint length occurred at the end of the operator closest to the *occQ* promoter, such that the footprint extended from -24 to -73 in the absence of octopine and extended from -35 to -73 in the presence of octopine. Third, octopine deceased the sensitivity of several DNase I-hypersensitive sites that were found interspersed between the protected residues. Fourth, and perhaps most striking, OccR incited a high angle DNA bend at its operator, and octopine relaxed the angle of this bend (Fig. 1, see ref. 34).

Hydroxyl radical footprinting experiments indicated that OccR binds to only one face of its operator (34). This face lies on the opposite side of the helix from the -35 motif, suggesting that OccR and RNAP can bind to this DNA simultaneously. Phasing assays were conducted to determine the direction of the DNA bend. We found that OccR bends operator DNA toward itself, and that it therefore occupies the interior angle to the DNA bend (34). Consequently, we believe that RNAP must occupy the exterior angle of this bend. If so, this would be unlike several other promoters that have protein-incited DNA bends, since RNAP generally binds to the interior angle of these bends (32).

A series of 3' and 5' resections of the OccR binding site was constructed by PCR amplification (35). These fragments were used to test the ability of OccR to bind and to cause ligand-responsive DNA bends. A 19 nucleotide fragment containing the upstream end of the wild type operator was found to be sufficient for high affinity OccR binding, provided that additional vector-derived sequences were also present in the region normally occupied by the remainder of the operator. Removal of any of this 19 nucleotide region (designated the High Affinity Subsite) abolished or drastically reduced OccR binding. This region contains an 8 nucleotide dyad symmetry (ATAA-N_7-TTAT, see Fig. 2). Three lines of evidence suggest that these nucleotides play a key role in OccR binding: First, the major grooves of these bases lie on the helical face that contacts OccR.

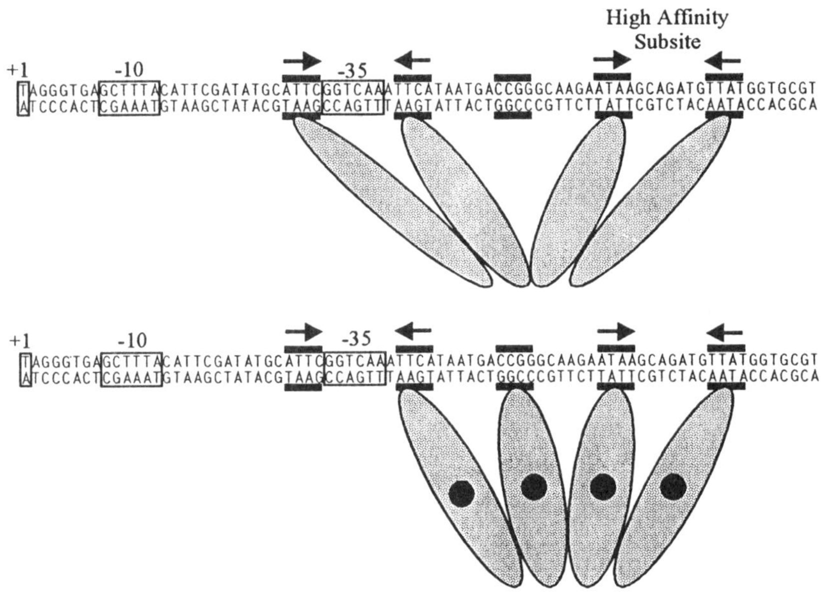

Figure 2. The sliding dimer model for OccR function. OccR binds tightly to a 19 nucleotide dyad symmetrical sequence denoted the High Affinity Subsite both in the presence and absence of octopine. In the absence of octopine, it contacts similar sequences flanking the -35 motif, while in the presence of octopine, two OccR monomers are thought to translocate by 10 nucleotides toward the high affinity subsite (Reprinted, by permission, from ref. 35).

Second, these sequences contain the T-N_{11}-A rule of virtually all LysR-type binding sites (16). Third, this sequence resembles the binding motifs of the homologous OxyR protein of *S. typhimurium* (29).

The downstream half of the OccR operator contains bases that show a weak similarity to the ATAA-N7-TTAT motif. Significantly, the major grooves of these bases are located on the helical face contacted by OccR. We propose that these nucleotides provide binding sites for OccR, and that the affinity of the protein for these sites is sufficiently weak to allow octopine-responsive sliding of the OccR. We propose that one dimer of OccR remains fixed at the high affinity subsite, while a second dimer of OccR can translocate between two sites 10 nucleotides apart (35, 34, see Fig. 1, 2).

The *occ* operon contains no fewer than 14 genes and is the longest known transcription unit on the Ti plasmid (Fig. 3). The first four genes (*occQ, occM, occP* and *occJ*) direct the uptake of octopine into the bacterial cell (31). The next three genes (*ooxB*, *ooxA*, and *ocd*) encode part of the octopine catabolic pathway, since *ooxB* and *ooxA* convert octopine into arginine and pyruvate, while *ocd* converts ornithine to proline (26, 38). The conversion of arginine to ornithine requires a gene (*arc*) located elsewhere (4, 3), although, interestingly, a fragment of the *arc* gene is located between *ooxA* and *ocd* (38), indicating that an ancestor of this Ti plasmid probably contained a functional *arc* gene, as is found at the corresponding position of nopaline-type Ti plamsids.

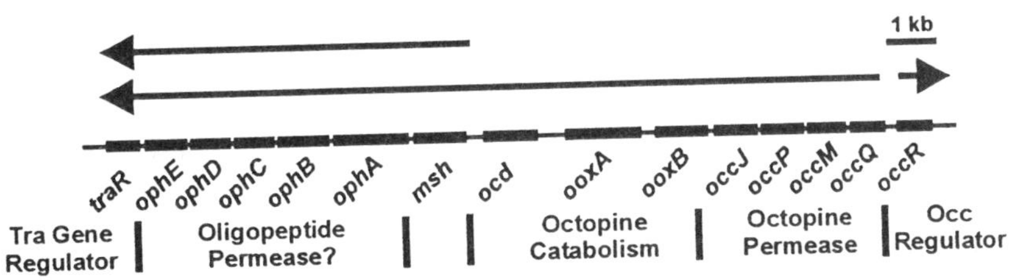

Figure 3. The *occ* operon. OccR activates a 14 gene operon containing two ABC-type permeases, three genes required for octopine catabolism, one gene (*msh*) of unknown function, and *traR*, a *tra* gene transcriptional activator. An internal promoter located between *ocd* and *msh* is activated by TraR.

Downstream of *ocd* lies a gene of unknown function (designated *msh*) followed by five genes encoding proteins homologous to the family of peptide permeases (14). While the products of these genes are presumed to transport some substrate, the identity of that substrate remains a mystery. The last gene in this operon is *traR*, which is a positive transcriptional regulator of the Ti plasmid *tra* regulon (10). As expected, OccR activates transcription of *traR* in response to octopine, fully explaining older observations that octopine is required for Ti plasmid conjugation in wild type strains (18).

TraR is believed to form complexes with an acyl homoserine lactone called *Agrobacterium* autoinducer, or AAI, whose synthesis is directed by the product of the *traI* gene (Fig. 4, see ref. 10). TraR-AAI complexes activate at least five promoters. First, they activate the divergent *traA* and *traC* promoters (Fig. 5, see ref. 13), and the *traI* promoter. The *traA*, *traC*, and *traI* promoters have conserved *tra* boxes that are essential for regulated transcription (Fig. 5, solid bar). By analogy to the nopaline-type Ti plamsid pTiC58 (5), the *oriT* is believed to be located directly downstream from the *traA* promoter (Fig. 5, hatched bars), and it remains an intriguing possibility that TraA, which probably binds to *oriT* (see below), could repress transcription of the *traA* promoter.

TraR-AAI complexes also activate a promoter that lies between *ocd* and *msh* (13). This results in a positive regulatory feedback loop such that activation of the *tra* regulon by TraR results in an increase in the pool size of TraR protein. This could have the effect of reinforcing the decision to induce *tra* gene expression. Activation of the *traI* gene should in principle create a second autoregulatory loop, since the *traI* product synthesizes AAI (see below). These positive autoregulatory loops are reminiscent of positive autoregulation of the *virA* and *virG* genes by phospho-VirG, which could have a similar positive reinforcement of induction of the *vir* regulon (37). The fifth promoter known to be activated by TraR directs the transcription of the *traM* gene, which encodes an antagonist of TraR (12). Activation of *traM* expression by TraR implies that *tra* gene expression is controlled by a negative feedback loop as well as the two positive feedback loops described above.

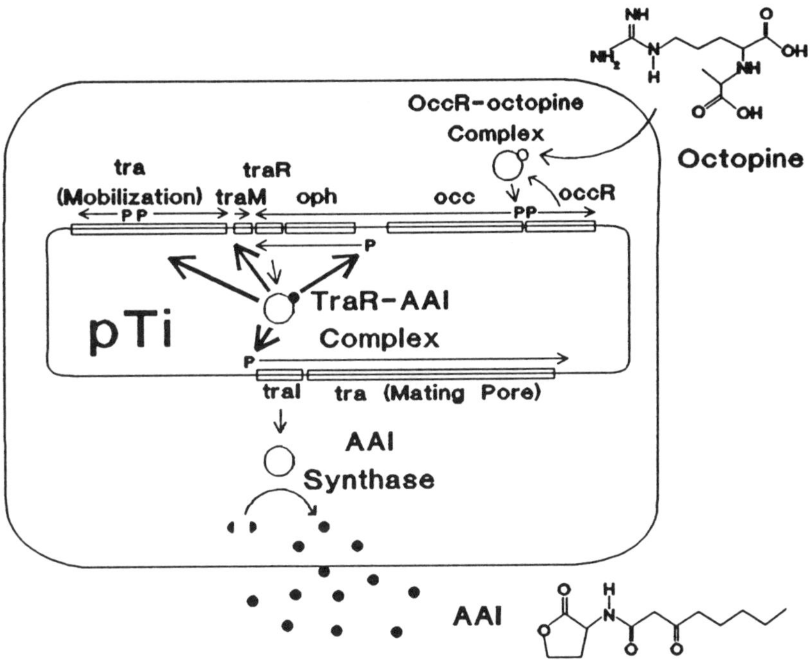

Figure 4. A model describing the regulation of Ti plasmid conjugal transfer genes. *traR* is activated by OccR in the presence of octopine. TraR forms complexes with AAI; these complexes activate five promoters. Since AAI is membrane-permeable, sufficient intracellular concentrations of it require high densities of conjugal donors (Reprinted, by permission, from ref. 10).

Although the *traR* and *traM* promoters are activated by TraR protein, they lack recognizable *tra* boxes.

TraI is homologous to a family of putative autoinducer synthases (11), none of which has been purified in an active form. In an effort to reconstitute TraI activity *in vitro*, we first fused it to a histidine tag, and found that the resulting fusion protein was active *in vivo*, indicating that its activity was unaffected by the fusion (23). This protein was solubilizied from inclusion bodies using guanidine HCl, and was applied onto a nickel affinity column in the presence of this

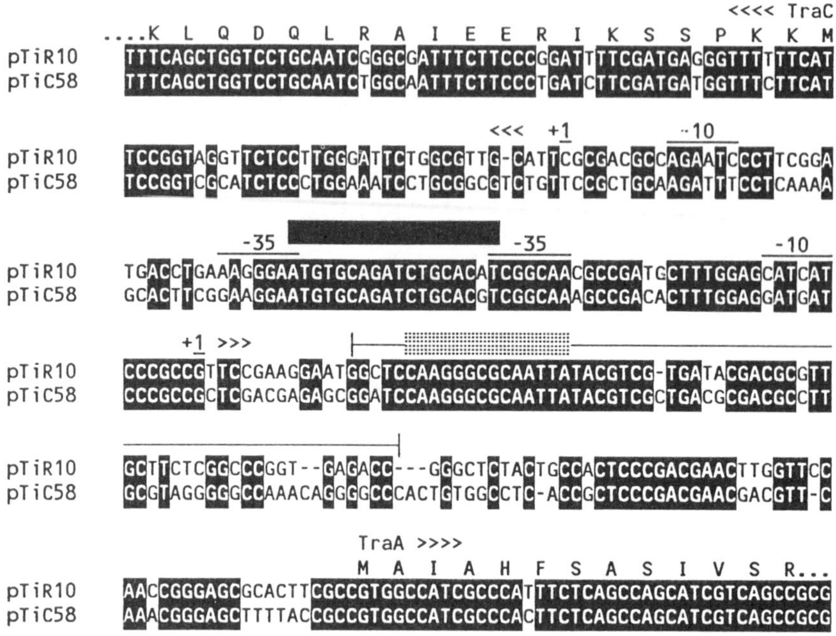

Figure 5. Sequence similarity between the octopine-type Ti plasmid pTiR10 and the nopaline-type Ti plasmid pTiC58 in the region containing the *traA* and *traC* promoters and *oriT* (hatched bar). *traA* and *traC* are activated by TraR bound to a single *tra* box (black bar).

denaturant. After gradual removal of denaturant, the protein was eluted in a soluble and active form by addition of imidazole (23).

The homoserine lactone moiety of AAI is believed to be obtained from S-adenosyl-methionine (AdoMet), while the 3-oxo-octanoyl moiety is believed to be obtained from 3-oxo-octanoyl-ACP (Fig. 6, see ref. 8). To test this, we incubated TraI with a crude extract of *E. coli* in the presence of AdoMet. This reaction produced a bioactive compound that coeluted with AAI by reverse phase HPLC (23). Addition of cerulenin, an inhibitor of fatty acid biosynthesis, abolished AAI production, suggesting that the fatty acid moiety is diverted from fatty acid biosynthesis intermediates. Dialyzed crude extract failed to support AAI production, indicating that one or more low molecular weight components are required (23). Activity was restored by addition of malonyl-CoA and NADPH, two substrates in fatty acid biosynthesis (to our knowledge, malonyl-CoA is used only in fatty acid biosynthetic reactions). These data provide

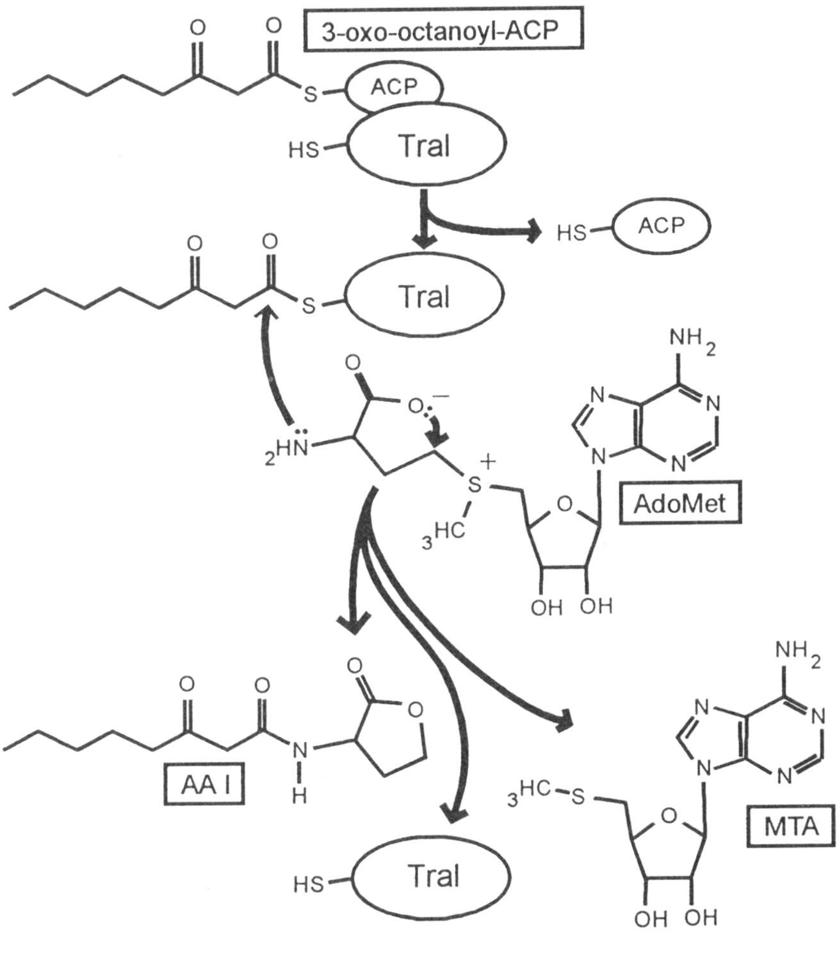

Figure 6. A model to describe AAI biosynthesis. According to this model, TraI acquires a 3-oxo-octanoyl groups from 3-oxo-octanoyl ACP. TraI catalyzes formation of the amide bond between the amino group of AdoMet and carbon 1 of the fatty acid, and also lactonizes Ado-Met to form the homoserine lactone moiety, releasing AAI and methyl-thio-adenosine (MTA).

further evidence that the fatty acid moiety of AAI is obtained from fatty acid biosynthetic intermediates.

To learn more about the genes regulated by TraR, we have determined the DNA sequence of the entire *tra* regulon and compared it to the sequences of other known *tra* genes (1). The putative mating bridge, encoded by 11 *trb* genes, is most strongly homologous to the *trb* operon of the IncP plasmid RP4 (22). All 11 RP4 *trb* genes have homologs in the Ti plasmid, and they are arranged in the same order except for one translocation, since, the *trbJ*, *trbK*, and *trbL* genes are promoter distal in RP4, but are located further upstream in the Ti plasmid. Some Ti plasmid *trb*-encoded proteins also resemble proteins encoded in the *virB* operon, but the resemblance is far less pronounced than to RP4 *trb*-encoded proteins. However, the *virB* operon closely resembles *tra* genes of the IncN plasmid pKM101 (25). This indicates that the two mating pores encoded on the Ti plasmid lie on different branches of the *tra* gene evolutionary dendrogram.

Interestingly, TrbK protein from the Ti plasmid is very weakly similar to its counterpart in RP4. TrbK of RP4 mediates entry exclusion (22), and in preliminary assays, we were unable to detect entry exclusion by the Ti plasmid. This could be significant, since Ti plasmid conjugal donors release AAI and therefore stimulate the *tra* regulon of other conjugal donors. It seems plausible therefore that donors may transfer the Ti plasmid to each other rather than to Ti plasmidless recipients. Efficient conjugation between donors would be prevented by entry exclusion.

A similar analysis of the conjugal DNA processing (*tra*) region results in a more complicated scheme. Two of these *tra*-encoded proteins most closely resemble Tra proteins of RP4. In contrast, no Ti plasmid *tra*-encoded protein resembles the RP4 conjugal nickase (TraI). Instead, sequence analysis predicts that an uncommonly large protein (TraA) is the nickase. TraA appears to have at least three domains, an amino terminal domain that is homologous to a family of conjugal nickases (including that of RSF1010), a central domain that is homologous to a family of conjugal helicases (including those found on IncN, IncW and IncF plasmids), and a carboxyl terminal domain that is not related to any known protein. This third domain has an internal in-frame initiation codon and is expressed as a separate protein as well as being expressed as a part of the larger TraA protein. We conclude that the Ti plasmid *tra* region evolved by appropriating *tra* genes from at least three sources: the TraA nickase from one group that includes RSF1010, the TraA helicase from a second group that includes IncN, IncW, and IncF plasmids (21, 30), and most other Tra and Trb proteins from an RP4-like plasmid.

Interestingly, the Ti plasmid *vir* region appears also to have chimeric origins, since the *virB*-encoded proteins most closely resemble proteins encoded in the IncN or IncW mating bridge operons, while proteins encoded by the *virD* operon most closely resemble Tra proteins of RP4 (1). Still other Vir proteins (including those encoded by the *virA*, *virC*, *virE*, *virG* and *virH* operons) do not have homologs in any characterized conjugation system. We have also observed a recent duplication of *tra* genes within the *vir* region. Extremely close homologs of the 3' end of *traA* and all of *traF* are found between *virE* and *virF* (unpublished data). It is therefore apparent that elements of the *vir* regulon have been adapted from many sources.

It has long been known that the widely studied pTiA6NC (which is the Ti plasmid of *A. tumefaciens* strain A348) is unable to transfer itself autonomously. This plasmid was also known to have a 12.65 kb deletion relative to several other octopine-type Ti plasmids (19, 27). We have found that this deletion removes most of the promoter-distal gene (*trbI*) of the *trb* cluster, which probably explains this conjugation deficiency (1). An extremely similar or identical deletion is found in the naturally occurring pTi89.10 (7), but is not found in pTiA6 itself, which is conjugation proficient. We speculate that pTiA6NC and pTi89.10 may be clonal in origin and that pTiA6NC and pTiA6 may be more distantly related.

Acknowledgments

We gratefully acknowledge support from the National Institute of General Medical Sciences, GM42893 (S.C.W.) and the United States Department of Agriculture, 9301084 (S.C.W.). C.F.

was supported by postdoctoral fellowship GM15128 from the National Institute of General Medical Science.

References

1. **Alt-Mörbe, J., J. L. Stryker, C. Fuqua, S. K. Farrand, and S. C. Winans.** 1996. The conjugal transfer system of *A. tumefaciens* octopine-type Ti plasmids is closely related to the transfer system of an IncP plasmid and distantly related to Ti plasmid *vir* genes. J. Bacteriol., **178**: 4248-4257.

2. **Beck von Bodman, S. B., J. E. McCatchan, and S. K. Farrand.** 1989. Characterization of conjugal transfer functions of *Agrobacterium tumefaciens* Ti plasmid pTiC58. J. Bacteriol. **171**: 5281-5289.

3. **Cho, K. C. Fuqua, B. S. Martin, and S. C. Winans.** 1996. Identification of *Agrobacterium tumefaciens* genes that direct the complete catabolism of octopine. J. Bacteriol. **178**: 1872-1880.

4. **Cho, K., C. Fuqua, and S. C. Winans.** 1996. Transcriptional regulation and location of *Agrobacterium tumefaciens* genes required for complete catabolism of octopine. J. Bacteriol., In press.

5. **Cook, D. M. and S. K. Farrand.** 1992. The *oriT* region of the Agrobacterium tumefaciens Ti plasmid pTiC58 shares DNA sequence identity with the transfer origins of RSF1010 and RK2/RP4 and with T-region borders. J. Bacteriol. **174**: 6238-6246.

6. **De Greve, H., H. Decraemer, J. Seurinck, M. van Montagu, and J. Schell.** 1981. The functional organization of the octopine *Agrobacterium tumefaciens* plasmid pTiB6S3. Plasmid **6**: 235-248.

7. **Dessaux, Y., J. Tempé, and S. K. Farrand.** 1987. Genetic analysis of mannityl opine catabolism in octopine-type *Agrobacterium tumefaciens* strain 15955. Mol. Gen. Genet. **208**: 301-308.

8. **Eberhard, A. T. Longin, C. A. Widrig, S. J. Stranick.** 1991. Synthesis of the *lux* gene autoinducer in *Vibrio fischeri* is positively autoregulated. Arch. Microbiol. **155**: 294-297.

9. **Farrand, S. K.** 1993. Conjugal transfer of *Agrobacterium* plasmids, p. 255-291. *In* D. B. Clewell (ed.), Bacterial Conjugation. Plenum Press, New York.

10. **Fuqua, C. and S. C. Winans.** 1996. Conserved cis-acting promoter elements are required for density-dependent transcription of *Agrobacterium tumefaciens* conjugal transfer genes. J. Bacteriol. **178**: 435-440.

11. **Fuqua, C. and S. C. Winans.** 1996. Localization of OccR-activated and TraR-activated promoters that express two ABC-type permeases and the *traR* gene of Ti plasmid pTiR10. Mol. Microbiol. (in press).

12. **Fuqua, C., M. Burbea, and S. C. Winans.** 1995. Activity of the *Agrobacterium* Ti plasmid conjugal transfer regulator TraR is inhibited by the product of the *traM* gene. J. Bacteriol. **177**: 1367-1373.

13. **Fuqua, W. C. and S. C. Winans.** 1994. A LuxR-LuxI type regulatory system activates *Agrobacterium* Ti plasmid conjugal transfer in the presence of a plant tumor metabolite. J. Bacteriol. **176**: 2796-2806.

14. **Fuqua, W. C., S. C. Winans, and E. P. Greenberg.** 1994. Quorum sensing in bacteria: the LuxR/LuxI family of cell density responsive transcriptional regulators. J. Bacteriol. **176**: 269-275.

15. **Gelvin, S. B., and L. L. Habeck.** 1990. *vir* genes influence conjugal transfer of the Ti plasmid of *Agrobacterium tumefaciens*. J. Bacteriol. **172**: 1600-1608.

16. **Goethals, K., M. M. Van, and M. Holsters.** 1992. Conserved motifs in a divergent nod box of *Azorhizobium caulinodans ors571* reveal a common structure in promoters regulated by LysR-type proteins. Proc. Nat. Acad. Sci., U.S.A. **89**: 1646-1650.

17. **Habeeb, L. F., L. Wang, and S. C. Winans.** 1991. Transcription of the octopine catabolism operon of the *Agrobacterium* tumor-inducing plasmid pTiA6 is activated by a LysR-type regulatory protein. Mol. Plant Microbe. Interact. **4**: 379-385.

18. **Klapwijk, P. M., T. Scheulderman, and R. Schilperoort.** 1978. Coordinated regulation of octopine degradation and conjugative transfer of Ti plasmids in *Agrobacterium tumefaciens*: evidence for a common regulatory gene and separate operons. J. Bacteriol. **136**:775-785.

19. **Knauf, V. C. and E. W. Nester.** 1982. Wide host range cloning vectors: a cosmid clone bank of an *Agrobacterium* Ti plasmid. Plasmid **8**:45-54.
20. **Lessl, M., D. Balzer, W. Pansegrau, and E. Lanka.** 1992. Sequence similarities between the RP4 Tra2 and the Ti VirB region strongly support the conjugation model for T-DNA transfer. J. Biol. Chem. **267**:20471-20480.
21. **Llosa, M., S. Bolland, and F. de la Cruz.** 1994. Genetic organization of the conjugal DNA processing region of the IncW plasmid R388. J. Mol. Biol. **235**:448-464.
22. **Lyras, D., A. W. S. Chan, J. McFarlane, and V. A. Stanisich.** 1994. The surface exclusion system of RP1: Investigation of the roles of *trbJ* and *trbK* in the surface exclusion, transfer, and slow-growth phenotypes. Plasmid **32**:254-261.
23. **Moré, M. I., L. D. Finger, J. L. Stryker, A. Eberhard, S. C. Winans.** 1996. Enzymatic synthesis of quorum-sensing autoinducer through use of defined substrates. Science, **272**: 1655-1658.
24. **Pansegrau, W., E. Lanka, P. T. Barth, D. H. Figurski, D. G. Guiney, D. Haas, D. R. Helinski, H. Schwab, V. A. Stanisich, and C. M. Thomas.** 1994. Complete nucleotide sequence of Birmingham IncP-alpha plasmids: Compilation and comparative analysis. J. Mol. Biol. **239**:623-663.
25. **Pohlman, R. F., H. D. Genetti, and S. C. Winans.** 1994. Common ancestry between IncN conjugal transfer genes and macromolecular export systems of plant and animal pathogens. Mol. Microbiol. **14**:655-668.
26. **Schindler, U., N. Sans, and J. Schröder.** 1989. Ornithine cyclodeaminase from octopine Ti plasmid Ach5: identification, DNA sequence, enzyme properties, and comparison with gene and enzyme from the Nopaline Ti plasmid C58. J. Bacteriol. **171:**847-854.
27. **Sciaky, D., A. L. Montoya, and M. D. Chilton.** 1978. Fingerprints of *Agrobacterium* Ti plasmids. Plasmid **1**:238-253.
28. **Steck, T. R. and C. I. Kado.** 1990. Virulence genes promote conjugative transfer of the Ti plasmid between *Agrobacterium* strains. J. Bacteriol. **172**:2191-2193.
29. **Toledano, M. B., I. Kulik, F. Trinh, P. T. Baird, T. D. Schneider, and G. Storz.** 1994. Redox-dependent shift of

OxyR-DNA contacts along an extended DNA binding site: a mechanism for differential promoter selection. Cell **78**:897-909.

30. **Traxler, B. A. and E. G. Minkley Jr.** 1987. Revised genetic map of the distal end of the F transfer operon: implications for DNA helicase I, nicking at *oriT*, and conjugal DNA transport. J. Bacteriol. **169**:3251-3259.

31. **Valdivia, R. H., L. Wang, and S. C. Winans.** 1991. Characterization of a putative periplasmic transport system for octopine accumulation encoded by *Agrobacterium tumefaciens* Ti plasmid pTiA6. J. Bacteriol. **173**:6398-6405.

32. **Van Der Vliet, P. C. and C. P. Verrijzer.** 1993. Bending of DNA by transcription factors. Bioessays **15**:25-32.

33. **Wang, L., J. D. Helmann, and S. C. Winans.** 1992. The *A. tumefaciens* transcriptional activator OccR causes a bend at a target promoter, which is partially relaxed by a plant tumor metabolite. Cell **69**:659-667.

34. **Wang, L. and S. C. Winans.** 1995. High angle and ligand-induced low angle DNA bends incited by OccR lie in the same plane with OccR bound to the interior angle. J. Mol. Biol. **253**:32-38.

35. **Wang, L. and S. C. Winans.** 1995. The sixty nucleotide OccR operator contains a subsite essential and sufficient for OccR binding and a second subsite required for ligand-responsive DNA bending. J. Mol. Biol. **253**:691-702.

36. **Winans, S. C.** 1992. Two-way chemical signaling in *Agrobacterium*-plant interactions. Microbiol. Rev. **56**:12-31.

37. **Winans, S. C., N. J. Mantis, C.-Y. Chen, C.-H. Chang., and D. C. Han**. 1994. Host recognition by the VirA, VirG two-component regulatory proteins of *Agrobacterium tumefaciens*. Research Microbiol. **145**:461-473.

38. **Zanker, H., G. Lurz, U. Langridge, P. Langridge, D. Kreusch, and J. Schröder.** 1994. Octopine and nopaline oxidases from Ti plasmids of *Agrobacterium tumefaciens*: molecular analysis, relationship, and functional characterization. J. Bacteriol. **176:**4511-4517.

CHAPTER 6

Organization and Assembly of the *Agrobacterium tumefaciens* T-Complex Transport Apparatus

Giulietta M. Spudich, Tu Anh Dang, David Fernandez, Xue-Rong Zhou, and Peter J. Christie*

Department of Microbiology and Molecular Genetics,
The University of Texas Health Science Center,
Houston, TX 77030

ABSTRACT: The *Agrobacterium tumefaciens virB* gene products are proposed to assemble into a transport system capable of exporting DNA across the bacterial envelope en route to bacterial and plant recipient cells. To gain information about the general organization of this DNA transport system, we have examined the topologies of the VirB proteins with a combination of approaches, including subcellular localization, fusion to periplasmically-active alkaline phosphatase, and protease susceptibility (11). We biochemically confirmed that VirB7 is a membrane-associated lipoprotein (11). By analysis of a collection of *virB* gene deletion derivatives of octopine strain A348, we gained evidence that VirB7, in conjunction with VirB9, promotes the assembly of a stabilized complex of VirB proteins (12). Finally, we discovered that VirB7 interacts directly with VirB9 via formation of intermolecular disulfide bridges between reactive Cys-24 and Cys-262 residues, respectively (34). Taken together, our findings support a model in which biosynthesis of VirB7 as a lipoprotein and formation of disulfide cross-linked VirB7 dimers represent critical early stages in the biogenesis of the T-complex transport apparatus.

Introduction

Agrobacterium tumefaciens incites tumor formation in susceptible plants by processing and exporting a nucleoprotein complex (T-complex) which most probably consists of single-stranded DNA (T-DNA), VirD2, a 5' capping protein, and

VirE2, a single-stranded DNA-binding protein (9, 20, 44). Early genetic studies suggested that the VirB proteins contribute to the transmission of T-complexes to plant cells (35). More recent sequence analyses have shown that the VirB proteins exhibit extensive sequence similarities to Tra proteins encoded by the broad-host-range (BHR) plasmids of the IncN (28), IncP (25), and IncW (see ref. 21) incompatibility groups. Furthermore, both the VirB and Tra transport systems of the BHR plasmids have been shown to promote the conjugal transfer of a common substrate, the non-selftransmissible IncQ plasmid, RSF1010 (3, 6). These findings led to the view that *A. tumefaciens* has adapted an interbacterial conjugal transport system for use in delivering oncogenes to plant cells (24, 43, 44). Even more intriguingly, components of each of these transport systems bear sequence similarities to components of the *Bordetella pertussis* Ptl transporter, which directs the export of the 6-subunit pertussis toxin to mammalian cells (42, 43). Thus, a family of transporters has been identified that collectively exhibits the ability to export macromolecules as diverse as nucleoprotein particles and multicomponent proteins across the gram-negative bacterial envelope en route to bacterial, plant, or mammalian recipient cells.

At present, little is known about the structures or assembly pathways of any of these transport systems. The investigations summarized here were focused in three areas: I) characterization of the general organization of the T-complex transport system, ii) identification of protein components required for assembly of a stabilized transport system, and iii) demonstration that two constituents of this transport apparatus are covalently cross-linked via intermolecular disulfide bridges. Implications of these findings for assembly of other macromolecular surface structures are discussed.

Results and Discussion

Localization of VirB proteins to the *A. tumefaciens* envelope. The general organization of the VirB components of the T-complex transport apparatus was characterized with a combination of three approaches, subcellular fractionation, proteinase K susceptibility, and construction of a collection of VirB::PhoA fusion proteins. Table 1 summarizes results of these studies.

Table 1. Properties of VirB proteins examined in this study[a]

VirB Protein	Subcellular[b] localization		Proteinase K[c] susceptibility			VirB/PhoA[d] Fusion Junction	PhoA[e] activity	
	IM	OM	WC	SP	LSP		-AS	+AS
VirB1	50	50	-	+	+	164	890	901
VirB2						121	1429	1298
VirB3						47	112	93
VirB4	90	10	-	+	+			
VirB5	60	40	-	+	+	93	1195	1130
VirB7	40	60	-	-	-	41	82	3070
B7::PhoA	30	70	-	+	+	41	82	3070
VirB8	90	10	-	+	+	72	439	411
VirB9	50	50	-	+	+	74	272	309
VirB10	50	50	-	-	+	276	974	1126
VirB11	70	30(C)[f]	-	-	+			

a. Data from Fernandez et al. (11).
b. Percentage of protein in inner (IM) and outer (OM) membrane fractions of a sucrose gradient.
c. Whole cells (WC), spheroplasts (SP), lysed spheroplasts (LSP).
d. Defined by the amino acid immediately preceding the fusion junction.
e. Expressed as Units of AP activity per OD_{600} in *A. tumefaciens* without (-) and with (+) acetosyringone (AS) induction.
f. VirB11 protein found in the cytoplasmic (c) fraction.

Cellular Fractionation: AS-induced A348 cells were fractionated into soluble (cytoplasm and periplasm) and insoluble (total membrane) material and examined for the presence of VirB proteins by SDS-PAGE and immunostaining of blots (11). Membrane material was further separated into cytoplasmic and outer membranes by two different techniques, isopycnic sucrose gradient centrifugation and Sarkosyl solubilization. Several of the VirB proteins were found to display a property unusual among membrane proteins by partitioning almost equally with the cytoplasmic and outer membrane fractions (11). This may be attributable to the assembly of the VirB proteins as a tight complex at fusion junctions between the two membranes (11). By contrast, the two ATPases, VirB4 and VirB11 (4, 10, 11, 30), and VirB8 (11) partitioned almost exclusively to the cytoplasmic membrane. VirB11 was unique among the VirB proteins examined in this study because a substantial proportion of protein (~30% of total protein) localizes to the cytoplasmic fraction (11, 30). In general, our findings agree with VirB localization data from other groups (8, 13, 14, 32, 38, 41).

Proteinase K susceptibility studies: The susceptibility of VirB proteins to proteinase K was tested with intact cells and spheroplasts of A348 (11). Protease treatment of whole cells caused no detectable reduction in levels of any of the VirB proteins examined. By contrast, treatment of spheroplasts with 100 or 200 μg/ml of proteinase K resulted in the complete or nearly complete degradation of VirB5, VirB8, VirB9, and VirB10. Furthermore, the disappearance of these proteins was not accompanied with a corresponding appearance of low-molecular-weight species reactive with antisera to these VirB proteins. This indicates that VirB5, VirB8, VirB9, and VirB10 reside predominantly in the periplasmic space and do not contain large cytoplasmic domains (11).

We found that protease treatment of whole cells, spheroplasts, and detergent-lysed spheroplasts caused little detectable loss of VirB7, indicating that VirB7 is intrinsically-resistant to digestion by proteinase K (11). Previous studies have shown that fusion junctions of chimeric PhoA proteins often are susceptible to protease digestion (see ref. 11). We demonstrated that this also is the case for the VirB7 derivative, VirB7::PhoA41; proteinase K treatment of spheroplasts but not whole cells expressing *virB7::phoA41* caused degradation of the fusion protein. These findings suggest that the VirB7 lipoprotein also resides in the periplasmic space (11).

Protease susceptibility studies of the two ATPases, VirB4 and VirB11, led to the important discovery that these proteins adopt different topologies at the cytoplasmic membrane. We found VirB11was completely resistant to proteases in whole cells or spheroplasts (11). This is consistent with other data from our lab (30) and other groups (13, 36) indicating that VirB11 is extrinsically-associated at the cytoplasmic face of the cytoplasmic membrane. By contrast, we found VirB4 was completely degraded in spheroplasts treated with 100 or 200 μg/ml of proteinase K (10). Furthermore, unlike the other VirB proteins examined in this study, degradation of VirB4 was accompanied by a corresponding appearance of two immunoreactive species of ~35 and ~45-kDa (10). This finding strongly suggests that VirB4 spans the cytoplasmic membrane at least once and contains two large, protease-inaccessible cytoplasmic domains. Further evidence that VirB4 possesses two potential periplasmic domains was provided with PhoA fusion studies (10).

PhoA fusion studies: To gain genetic evidence for the export of VirB proteins across the cytoplasmic membrane, we introduced *phoA*, which encodes periplasmically-active alkaline phosphatase (AP), into available restriction sites at positions in *virB1* through *virB3*, *virB5*, and *virB7* through *virB10* with a high probability of encoding an exported domain as determined by sequence analysis (22, 33, 37, 39, 40) Using a novel approach, we also constructed a saturated collection of *virB4::phoA* and *virB5::phoA* fusions (10). We quantitated the AP activities of a collection of VirB::PhoA fusions in *E. coli* and in uninduced and AS-induced *A. tumefaciens*. Table 1 lists the positions of the VirB/PhoA fusion junctions and the corresponding AP activities in uninduced and AS-induced A348 merodiploids. The quantitative assays showed that each of the fusion proteins generated in this study exhibited intermediate (50 - 200 units per OD_{600} unit) to high (> 200 units per OD_{600} unit) levels of AP activity (11). Both *E. coli* CC118 and *A. tumefaciens* A348 host cells reproducibly possessed background (0 - 2 units per OD_{600} unit) levels of AP activities. These findings provided genetic evidence that each of the VirB proteins, VirB1, VirB2, VirB3, VirB5 and VirB7 through VirB10 contain exported domains (11). VirB4::PhoA fusions with AP activities possessed junctions that mapped to two positions, one near the amino terminus and a second located immediately downstream of the ATP-binding domain (10).

During the course of these investigations, we observed that one VirB::PhoA fusion protein, VirB7::PhoA41, displayed an unusual property. Cells expressing *virB7::phoA41* exhibited a relatively low level of AP activity when incubated in induction media without acetosyringone (AS). By contrast, these cells exhibited a dramatic ~35-fold enhancement in AP activity when incubated in induction media containing AS. VirB7::PhoA41 protein content also increased substantially in AS-induced cells, as shown by development of immunoblots with anti-VirB7 or anti-PhoA antisera. We have shown that AS-dependent accumulation of VirB7::PhoA41 does not result from an increase in transcription of *virB7::phoA41.* Currently, we are testing whether VirB7::PhoA41 is stabilized through an interaction with another Vir protein whose synthesis is dependent on VirA/VirG activation of gene expression.

Based on results of these investigations, we depict the T-complex transport apparatus as a complex of VirB proteins consisting of VirB4 and VirB11 ATPases localized at the cytoplasmic face of the cytoplasmic membrane, VirB6, a multispanning cytoplasmic membrane protein, and VirB5, VirB7, VirB8, VirB9, and VirB10 positioned predominantly or exclusively within the periplasmic space possibly as a structure extending from the cytoplasmic membrane to the outer membrane (Figure 1) (11). VirB1, VirB2, and VirB3 may also localize in the periplasmic space, or at the cell exterior possibly assembled as a pilus structure (11, 32). This transport system likely also consists of other Vir protein components such as VirD4 (3).

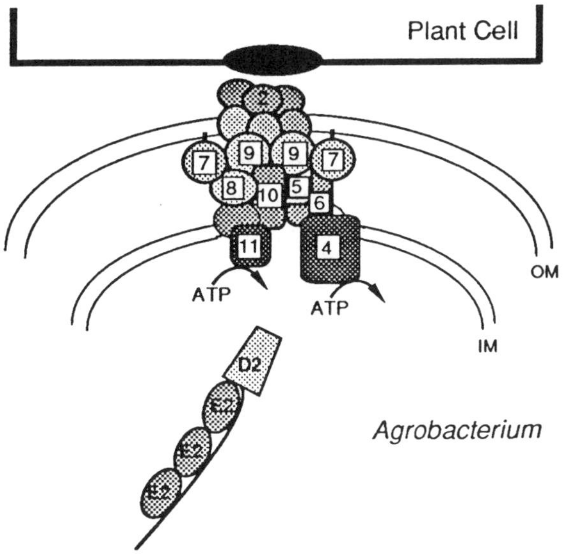

Figure 1. Model of the general organization of the T-complex transport apparatus.

VirB7 and VirB9 promote accumulation of VirB proteins. To identify the essential VirB proteins, we constructed a set of 11 A348 derivatives sustaining precise *virB* gene deletions (4, 5). Virulence assays showed *virB2* through *virB11* encode essential Vir proteins, whereas *virB1* encodes a function that is important but not essential for *A. tumefaciens* infection of dicot plants (5). Furthermore, each of the Δ*virB* mutations could be complemented with plasmids expressing the corresponding *virB* genes, demonstrating that each of the precise deletions is nonpolar on expression of downstream genes (5). However, immunoblot studies of these strains showed that several of the Δ*virB* mutants, notably the Δ*virB7* and Δ*virB9* mutants, possessed diminished steady-state levels of several VirB proteins (5). These findings raised the possibility that certain VirB proteins such as VirB7 and VirB9 function as stabilizing elements during assembly of the T-complex transport apparatus.

We developed two lines of study to test for a correlation between synthesis of VirB7 and VirB9 and accumulation of other VirB proteins. The first involved an analysis of VirB protein levels in

A348 derivatives sustaining precise deletions of *virB7*, *virB8*, and *virB9*, now designated strains PC1007, PC1008, and PC1009, respectively.

Table 2 summarizes the principal findings from these studies. A precise deletion of *virB7* was correlated with a significant reduction in steady-state levels of VirB proteins (5, 12). The *trans* expression of *virB7* in PC1007 cells partially restored levels of these proteins and

Table 2. Abundance of VirB proteins in various Δ*virB* mutants[a]

Strain[b]	Plasmid[c] (*virB* genes expressed)	VirB Protein Abundance[d] -	+	+++	WT
PC1007	-	7,8,9	4,10	5,11	
	pPCB974 (*virB7*)	8	4,9,10	5,11	7
	pPCB975 (*virB7*, *virB8*)				4,5,7,8,9, 10,11
	pPCB985 (*virB8*)	7,9	4,10	11	5,8
	pED9 (*virB9*,*virB10* *virB11*)	7,8,9		4,10	5,11
PC1008	-	8		4	5,7,9,10,11
	pPCB985 (*virB8*)			4	5,7,8,9,10, 11
PC1009	-	8[e],9	4,10	5,11	7
	pPCB992 (*virB9*)	8[e]		4,5	7,9,10,11

a. Data from Fernandez et al. (12).

b. A348 derivatives precisely deleted of *virB7* (PC1007), *virB8* (PC1008), and *virB9* (PC1009).

c. IncP plasmids expressing the *virB* genes listed in parantheses. See ref. 12.

d. Relative amounts of VirB proteins from none detected (-), low (+) and moderate (+++) levels, and wild-type (WT) levels, as determined by densitometric scanning of immunoblots.

e. No VirB8 detected in cells harvested at 18 h post-induction. Small amounts of VirB8 were detected at 8 and 12 h post-induction.

trans coexpression of *virB7* and *virB8* fully restored levels of these proteins to wild-type levels (12). These data provided evidence that VirB7 promotes accumulation of other VirB proteins. Furthermore, VirB8 appears to play a role in stabilizing other VirB proteins in this genetic background. Interestingly, we determined that PC1008 cells accumulate abundant levels of most VirB proteins examined, except for VirB4 which is present at a reduced level (12). This latter finding suggests that VirB8 is nonessential for assembly of a stabilized complex of VirB proteins. At this point, we do not know why VirB8 appears to have a stabilizing effect on other VirB proteins in PC1007(pPCB975) cells. We note that pPCB975 constitutively expresses VirB7 and VirB8. Furthermore, we have determined that the Δ*virB7* mutation abolishes *virB8* expression and also has a slight polar effect on *virB9* expression (12). Therefore, it is possible that preexisting pools of VirB8 somehow augment a stabilizing activity of VirB7 under conditions in which the kinetics of synthesis or the pool sizes of other VirB proteins is perturbed.

A precise deletion of *virB9* was correlated with a significant reduction in steady-state levels of all the VirB proteins examined except for VirB7 which accumulated to wild-type levels (5, 12). Expression of *virB9* in *trans* restored the abundance of most VirB proteins to wild-type levels (12). These data established first that VirB9 directly contributes to the accumulation of other VirB proteins, and second that VirB7 synthesis in the absence of VirB9 fails to promote accumulation of VirB proteins.

Modulation of VirB proteins by regulated synthesis of VirB7. Results of these immunoblot studies led us to investigate whether VirB protein levels could be influenced by modulating expression levels of *virB7* (12). We tested for corresponding changes in levels of VirB7 and other VirB proteins in the following cell types: I) a Δ*virB7* mutant expressing *virB7* and *virB8* from isopropyl-β-D-thiogalactopyranoside (IPTG)-inducible P_{lac} and other *virB* genes from acetosyringone (AS)-inducible P_{virB}; ii) a Δ*virB* operon mutant expressing *virB7* and *virB8* from P_{lac} and *virB9*, *virB10*, and *virB11* from P_{virB}; and iii) a Δ*virB* operon mutant expressing *virB7* from IPTG-inducible P_{lac} and *virB9* from P_{virB}(12). To achieve regulatable expression of *virB7*, we introduced the *lacI*^{q} gene from pMMB22 onto pPCB975. The resulting plasmid, pPCB976, was ligated to: i) pSW172 (making pPCB976), ii) pED9 such that expression of *virB9 - virB11* from P_{virB} was unaffected (making pPCB115) or iii) pED9 at a site within *virB10* such that only *virB9* was expressed from P_{virB} (making pPCB970) (12). These cells were grown in increasing concentrations of IPTG and a constant amount of AS [200 µM for PC1007(pPCB976) and 20 µM for PC1000(pPCB115) and PC1000(pPCB970)]. The abundance of VirB proteins in the corresponding protein extracts was assessed by SDS-PAGE and immunoblot analysis (12).

Figure 2 shows that PC1007(pPCB976) accumulated VirB7 at a detectable level upon IPTG-induction with ≥20 µM IPTG. Steady-state levels then increased with apparent linearity over a range of 20 µM to 1 mM IPTG. Similar results were obtained for VirB8. By contrast, levels of VirE2 were unaffected by growth of cells in this range of IPTG. PC1007(pPCB976) cells induced with 1mM IPTG possessed VirB7 at wild-type levels and VirB8 at ~75 % of wild-type levels. The effect of modulating VirB7 synthesis on levels of other VirB proteins is summarized in Table 3. First, synthesis of VirB7, along with VirB8, promotes the accumulation of VirB4, VirB9, VirB10, and VirB11 in PC1007(pPCB976) cells. Second, synthesis of VirB7 promotes the accumulation of a subset of VirB proteins, including VirB7, VirB9, VirB10, and VirB11 in PC1000(pPCB115) cells in the absence of other VirB proteins. Third, synthesis of VirB7

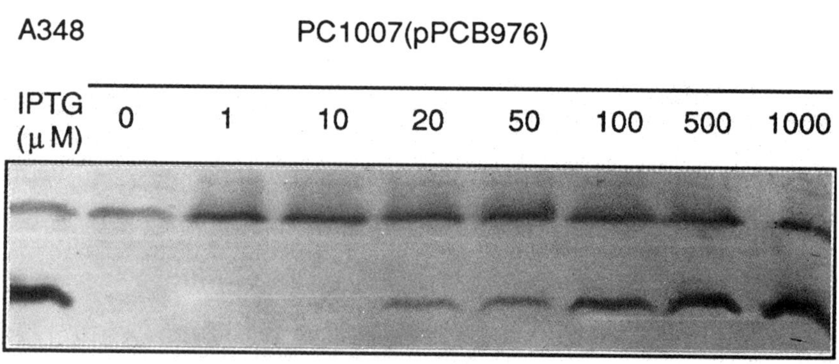

Figure 2. Accumulation of VirB7 in PC1007(pPCB976) as a function of [IPTG].

Table 3. Effects of modulating VirB7 synthesis on levels of other VirB proteins[a]

Strain	Description	Effects of increased VirB7 synthesis on levels of other VirB proteins[b]
PC1007 (pPCB976)	Δ*virB7*, *virB7, virB8* from P_{lac}, *lacI*^{q}	Increases in levels of VirB4, VirB8, VirB9, VirB10, VirB11 No increase in VirB5
PC1000 (pPCB115)	Δ*virB* operon[c] *virB7, virB8* from P_{lac}, *lacI*^{q} *virB9*, *virB10*, virB11 from P_{virB}	Increases in levels of Vir B9, VirB10, VirB11 No increase in VirB8
PC1000 (pPCB970)	Δ*virB* operon *virB7* from P_{lac}, *lacI*^{q} *virB9* from P_{virB}	Increases in levels of VirB9

a. Data summarized from Fernandez et al. (12).
b. Cells were grown in 0, 20, 100, and 1000 μM IPTG.
VirB protein content was examined by SDS-PAGE and immunostaining.
c. An A348 derivative sustaining a precise deletion of the entire *virB* operon (11).

promotes the accumulation of VirB9 in PC1000(pCPB970) cells which are devoid of other VirB proteins (12). Together, our data form the basis of two predictions. First, VirB7 promotes the assembly of a stabilized complex of VirB7, VirB9, VirB10, and VirB11. Second, VirB7 stabilizes VirB9 as a result of a direct protein - protein interaction (12).

VirB7 forms two intermolecular disulfide bridges. We attempted to demonstrate a direct interaction between VirB7 and VirB9 with the use of chemical cross-linking reagents. During the course of these investigations, we discovered that VirB7 and VirB9 assemble as a heterodimer that withstands exposure to detergents, NaCl, urea, and electrophoresis through SDS-polyacrylamide gels. Furthermore, the heterodimer was dissociated by treatment of cell extracts with the reducing agents, 2-mercaptoethanol or dithiothreitol. Three lines of study confirmed that VirB7 forms intermolecular disulfide bridges.

First, we assayed protein extracts electrophoresed under nonreducing conditions for the presence of high molecular weight protein species reactive with VirB7 and VirB9 antisera. We found that nonreduced extracts from PC1000(pPCB970) cells possessed a 36-kDa species that was reactive with both anti-VirB7 and anti-VirB9 antisera. This is precisely the size expected of a heterodimer comprised of VirB7 (~4.5-kDa) and VirB9 (32-kDa). Furthermore, we determined that the amount of the 36-kDa species, designated complex C1, increased as a function of the concentration of IPTG added to cell cultures for induction of *virB7* expression (12). A second species, also crossreactive with both antisera, was detected at ~18-20 kDa. This species, designated complex C2, also increased as a function of [IPTG] added to PC1000(pPCB970) cultures.

Second, we assessed whether anti-VirB7 and anti-VirB9 antisera would co-precipitate the presumptive VirB7/VirB9 heterodimer from cell extracts suspended in nonreducing buffers. Immunoblot analyses of these immunoprecipitates showed the presence of complex C1, the presumptive VirB7/VirB9 heterodimer (34). Thus, anti-VirB7 and anti-VirB9 antisera co-precipitated the same species. Interestingly, we discovered that we could cause dissociation of complex C1 into the VirB7 and VirB9 monomers by electrophoresis of nonreduced immunoprecipitates adjacent to a lane containing 2-mercaptoethanol. Using glycine SDS/(12.5%) polyacrylamide gels, the 2-mercaptoethanol diffused halfway into the adjacent lanes, causing dissociation of the presumptive VirB7/VirB9 heterodimer (34). In addition, analysis of the immunoprecipitates with a tricine-SDS/polyacrylamide gel system revealed that VirB7 also forms a dissociable homodimer. We designated this as complex C3 (34).

Third, we examined the composition of the presumptive VirB7/VirB9 heterodimer with two-dimensional nonreducing/reducing gel electrophoresis (see Figure 3). Total cellular protein from AS-induced A348 was loaded in 4 lanes of an

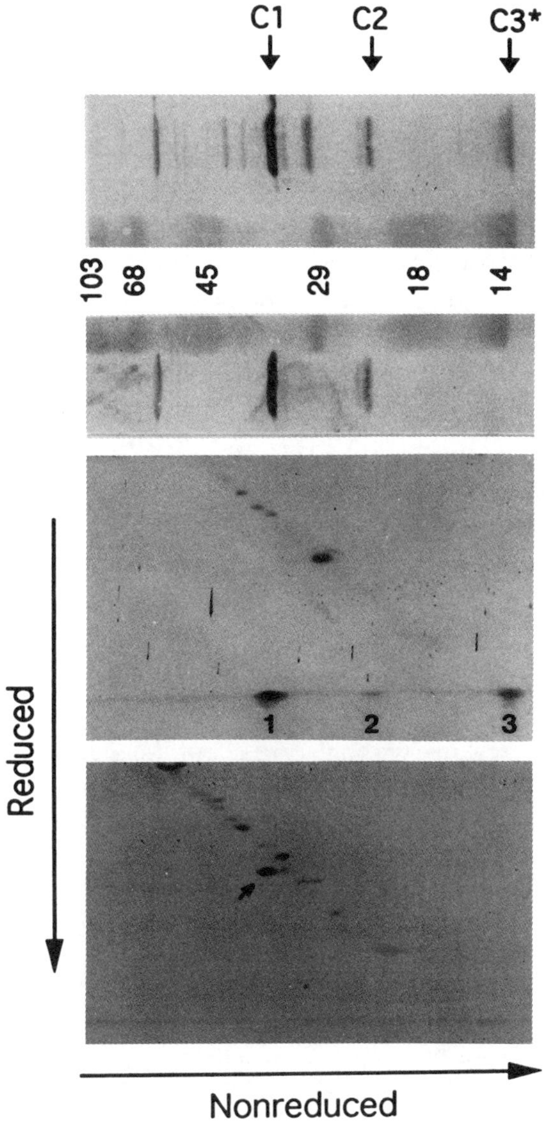

Figure 3. Two/dimensional nonreducing/reducing SDS/polyacrylamide gel electrophoresis of A348 extracts, as described in the text.

SDS-12.5% polyacrylamide gel and electrophoresed under nonreducing conditions. Proteins in two lanes were transferred to nitrocellulose and developed with VirB7 (top panel) or VirB9 (second panel) antibodies to mark the positions of complexes C1, C2, and C3. The second two lanes were reduced by soaking in 10 mM 2-mercaptoethanol before electrophoresis in the second dimension. Material was transferred to nitrocellulose and developed with either VirB7 (third panel) or VirB9 (bottom panel) antibodies. A diagonal was generated by crossreactivity of the antibodies with monomeric proteins or nonreducible complexes. Complex C1, the presumptive VirB7/VirB9 heterodimer, was not located on this diagonal; instead, two dissociation products, VirB7 (spot 1) and VirB9 (marked by arrow), were immunologically detected. This confirms that the reducible C1 complex is a VirB7/VirB9 heterodimer. Complex C2 was reactive with both VirB7 and VirB9 antisera in the one-dimensional gels, and dissociated into a VirB7 (spot 2) monomer upon exposure to 2-mercaptoethanol. However, the VirB9 monomer was not immunologically detected with the two-dimensional gels. This could be attributable to the relatively low abundance of complex C2 coupled with the fact that the titer of our VirB9 antisera is lower than that of the VirB7 antisera. Alternatively, complex C2 is not comprised of VirB9. Complex C3, the presumptive VirB7 homodimer, was reactive with anti-VirB7 but not anti-VirB9 antisera. This complex migrated with the dye front of the glycine SDS/polyacrylamide gel. In other experiments, we showed that complex C3 corresponds to a VirB7 homodimer.

Disulfide-crosslinking occurs *in vivo*. To confirm that the dissociable complexes identified in these studies are formed *in vivo*, we routinely treated cells with the thiol alkylating agent, *N*-ethylmaleimide (NEM), which blocks free SH groups and prevents the formation of disulfide bonds (31). Addition of NEM at 10 mM to A348 cultures immediately prior to cell pelleting and boiling in sample buffer had no effect on formation of complexes C1, C2 or C3. We also established that none of the complexes identified was affected by addition of 6 M urea to the samples prior to boiling. Based on these results, we conclude that the observed complexes form *in vivo* via the formation of disulfide bonds between reactive cysteine residues present in VirB7 and VirB9.

Cys-24 of VirB7 and Cys-262 of VirB9 are reactive cysteine residues. We identified the reactive cysteine residues of VirB7 and VirB9 by site-directed mutagenesis. VirB7 from pTiA6NC possesses four cysteines (39, 40), two in its cleaved signal sequence, an amino terminal Cys-15 residue which is predicted to be unavailable for intermolecular cross-linking (11, 19), and an internal Cys-24 residue. VirB9 possesses two cysteines (39, 40), one in its cleaved signal sequence and Cys-262. We examined the effect of Cys-24 and Cys-262 to Ser substitutions on the abilities of VirB7 and VirB9 to form dissociable complexes. Interestingly, the C24S and C262S derivatives failed to accumulate to detectable levels, suggesting that disulfide mediated dimerization is essential for stabilization of VirB7 and VirB9 (34). To further show that these cysteines participated in the formation of intermolecular disulfide bonds, we stabilized these proteins by fusing them to alkaline phosphatase. In initial studies, we found that VirB7::PhoA41 forms disulfide cross-linked homodimers and VirB7/VirB7::PhoA41 mixed dimers. Furthermore, we found that VirB9::PhoA293 forms disulfide cross-linked VirB7/VirB9::PhoA293 mixed dimers (34). By contrast, the stabilized derivatives, VirB7C24S::PhoA41 and VirB9C262S::PhoA293, failed to form any cross-linked dimers. Together, results of these studies established: i) VirB7 and VirB9 form intermolecular disulfide bridges via reactive Cys-24 and Cys-262 residues, respectively, and ii) Dimerization is essential for stabilization of both proteins (34).

A model for assembly of the T-complex transport system. Based on our studies, we propose that biogenesis of the T-complex transport system proceeds according to the sequence of events depicted in Fig. 4.

In stage I, VirB7 is processed as a lipoprotein, most probably according to the lipoprotein biosynthetic pathway developed for Lpp, the major outer membrane lipoprotein of *E. coli* (19, 26, 29). In stage II, newly-processed VirB7 monomers form intermolecular disulfide

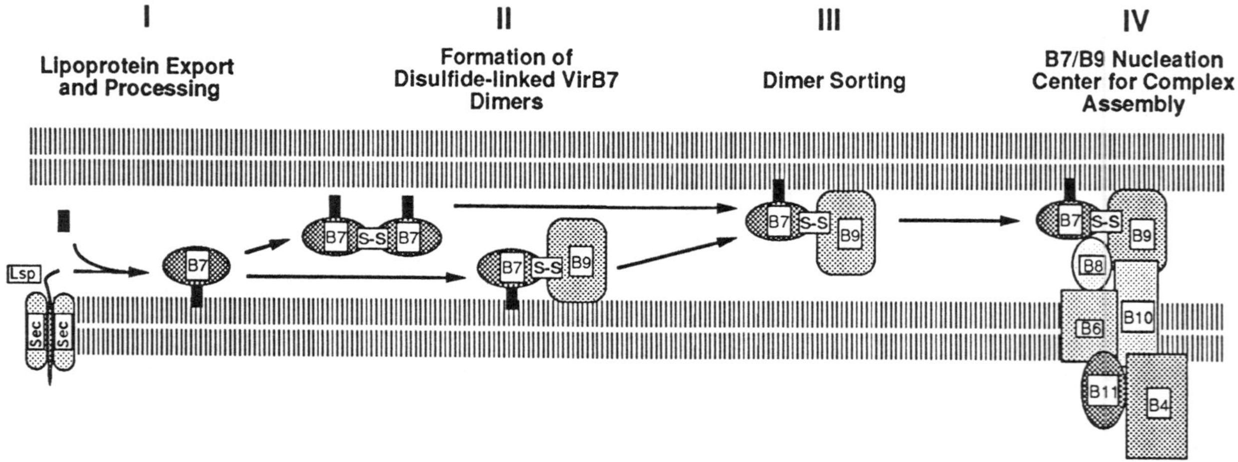

Figure 4. A model for the assembly of the T-complex transport system.

cross-links. As depicted in Fig. 4, we envision that this disulfide bond formation pathway proceeds along one of two routes. One possibility is that a thiol/disulfide interchange protein present in the periplasm transfers its reactive disulfide to form the VirB7 homodimer, then a disulfide isomerase catalyzes an intermolecular isomerization reaction that shuffles the disulfide bond of the VirB7 homodimer to form the stabilized VirB7/VirB9 heterodimer. A second possibility is that an oxidoreductase(s) catalyzes formation of both the VirB7 homodimers and VirB7/VirB9 heterodimers directly from newly-secreted monomers. For both pathways, the enzymes responsible for catalyzing formation of intermolecular disulfide bonds likely are analogs of the Dsb (Disulfide bond formation) proteins recently identified in *E. coli* and other species (1). These thiol/disulfide interchange proteins are known to catalyze the formation of intramolecular disulfide bonds in a wide variety of exported proteins, including secreted exotoxins as well as components of several eubacterial macromolecular surface structures such as *E. coli* flagella, *Vibrio cholerae* toxin-coregulated colonization pilus (TCP), and the *Haemophilus influenzae* competence system (see ref. 1).

In stage III, VirB7 dimers are sorted to their final destinations in the *A. tumefaciens* envelope. Based on the observed instabilities of the VirB7 and VirB9 monomers (34), we propose that dimer formation must precede or occur simultaneously with sorting. Thus, the VirB7 sorting pathway must accommodate three features that distinguish VirB7 from other known eubacterial lipoproteins (15, 19, 26, 29): i) VirB7 requires dimer formation for its stabilization (34), ii) dimerization results from formation of covalent cross-links (34), and iii) VirB7 co-partitions to both the cytoplasmic and outer membranes (11). It is tempting to speculate that the lipoprotein biosynthetic and disulfide bond formation pathways converge at the step of sorting VirB7 to its final destination. This could be achieved with a bifunctional periplasmic carrier protein that both catalyzes the formation of the observed intermolecular disulfide bonds and delivers the resulting, stabilized dimers to the appropriate membrane receptor (34). An analogous function has been suggested for TcpG, a DsbA homolog required for assembly of toxin-coregulated colonization pili (TCP) in *Vibrio cholera* (27).

Given the importance of VirB7 and VirB9 for stabilization of other VirB proteins (12), we propose that the VirB7/VirB9 heterodimer functions as a nucleation center for recruitment of other VirB proteins for further assembly of the T-complex transport apparatus (Stage IV). We have gained some insights into possible stabilizing contacts between the VirB7/VirB9 heterodimer and other VirB proteins. For example, *A. tumefaciens* cells expressing *virB7* and *virB8* from an IPTG-inducible *lac* promoter and *virB9*, *virB10*, and *virB11* from the AS-inducible *virB* promoter accumulate VirB9, VirB10, and VirB11 in rough proportion to the level of VirB7 synthesis (12), suggesting that the VirB7/VirB9 heterodimer interacts with VirB10 and VirB11. This is consistent with a report by Beaupre and Binns (2) that synthesis of VirB9 influences the oligomeric state of VirB10. We have shown that VirB7, VirB9, and VirB10 are membrane-associated proteins with large periplasmic domains (11). Thus, these proteins likely form stabilizing contacts in the periplasm. By contrast, as noted above, both VirB11 and VirB4, whose steady-state levels also are enhanced by VirB7 synthesis (12), are cytoplasmic-membrane bound proteins that reside predominantly or exclusively within the cell interior. This suggests that the stabilizing effect of the VirB7/VirB9 heterodimer on these two ATPases is mediated through another membrane-spanning VirB protein. This might be VirB10 (34), a monotopic protein with an amino-terminal cytoplasmic membrane-spanning domain (41).

Summary. In summary, our investigations have led to the identification of critical early reactions that must occur for assembly of a stabilized and functional T-complex transport apparatus. Are features of the proposed assembly pathway depicted in Figure 4 important for assembly of other conjugal DNA transport systems, or, more broadly, other macromolecular surface structures? It may be highly significant that lipoproteins have been shown to be associated with many eubacterial surface structures (see ref. 12, 19, 29). Their unique mode of attachment to cell membranes, coupled with data indicating that lipoproteins are sorted to specific membrane receptors (7, 26), is at least consistent with the possibility that these molecules serve as anchoring sites for biogenesis of macromolecular surface structures. What role intermolecular disulfide bonds play in any of these assembly processes remains to be determined. There is one dramatic example of intermolecular disulfide bond formation in eubacteria. The *Chlamydia* EnvA lipoprotein forms an extensive disulfide cross-linked supramolecular complex with other envelope proteins; this complex is thought to functionally-substitute for a missing peptidoglycan layer in these bacteria (18). Most other examples of intermolecular bridges identified in eubacteria have been shown to form between Cys-substituted derivatives of proteins that function as noncovalent dimers. Two examples include an A48C mutant derivative of Lpp, the major outer membrane lipoprotein of *E. coli* (16), and derivatives of the *E. coli* Trg chemoreceptor generated by cysteine-scanning mutagenesis (23). This raises the possibility that VirB7 acquired the ability to form disulfide cross-links as the result of a fortuitous event - the incorporation of a Cys residue by mutation - during its evolution from a progenitor lipoprotein. Consistent with this possibility, it is interesting to note that even though both VirB7 proteins from pTiA6NC (39, 40) and pTiC58 (22, 33) and possible functional analogs of VirB7, i.e., PtlI from *Bordetella pertussis* (43) and TrbH from the IncP plasmid, RP4αa (25), possess potentially-reactive internal cysteine residues, other VirB7 homologs, i.e., the TraN protein from the IncN plasmid, pKM101 (28) are devoid of internal cysteine residues. The observed sequence similarities of components of the VirB, Ptl, and BHR Tra transport systems and the functional-relatedness of the VirB and BHR Tra systems (3, 6, 17, 21, 25, 28, 43), strongly suggests that each of the VirB7-like proteins supplies a conserved function with respect to the assembly/function of the cognate transporter. If so, these lipoproteins play similar roles when assembled as covalent dimers, or as noncovalent oligomers or monomers.

Acknowledgments

We thank Drs. Anath Das and Christian Baron for also discussing their findings that VirB7 forms disulfide cross-linked dimers at this meeting. This work was supported by Public Health Service Grant GM48746.

References

1. **Bardwell, J. C. A.** 1994. Building bridges: disulfide bond formation in the cell. Molec. Microbiol. **14**: 199-205.
2. **Beaupre, C. E., and A. N. Binns.** 1994. Abstracts of the Fifteenth Annual Crown Gall Meeting. Cornell Univ., Ithaca, NY
3. **Beijersbergen, A., A. D. Dulk-Ras, R. A Schilperoort, and P. J. J. Hooykaas.** 1992. Conjugative transfer by the virulence system of *Agrobacterium tumefaciens.* Science **256:** 1324-1327.
4. **Berger, B. R., and P. J. Christie.** 1993. The *Agrobacterium tumefaciens virB4* gene product is an essential virulence protein requiring an intact nucleoside triphosphate-binding domain. J. Bacteriol. **175:** 1723-1734.
5. **Berger, B. R., and P. J. Christie.** 1994. Genetic complementation analysis of the *Agrobacterium tumefaciens virB* operon: *virB2* through *virB11* are essential virulence genes. J. Bacteriol. **176:** 3646-3660.
6. **Buchanan-Wollaston, V., J. E. Passiatore, and F. Cannon.** 1987. The *mob* and *oriT* mobilization functions of a bacterial plasmid promote its transfer to plants. Nature **328:** 172-175.
7. **Choi, D.-S., H. Yamada, T. Mizumo, and S. Mizushima.** 1986. Trimeric structure and localization of the major lipoprotein in the cell surface of *Escherichia coli.* J. Biol. Chem. **261:** 8953-8957.
8. **Christie, P. J., J. E. Ward, M. P. Gordon, and E. W. Nester.** 1989. A gene required for transfer of T-DNA to plants encodes an ATPase with autophosphorylating activity. Proc. Natl. Acad. Sci. USA **86:** 9677-9681.
9. **Christie, P. J., J. E. Ward, S. C. Winans, and E. W. Nester.** 1988. The *Agrobacterium tumefaciens virE2* gene product is a single-stranded DNA binding protein that associates with T-DNA. J. Bacteriol. **170:** 2659-2667.
10. **Dang, T. A., and P. J. Christie**. 1996. Unpublished data.
11. **Fernandez, D., T. A. Dang, G. M. Spudich, X.-R. Zhou, B. R. Berger and P. J. Christie.** 1996. *Agrobacterium tumefaciens virB7* gene product, a proposed component of the T-complex transport apparatus, is a membrane-associated lipoprotein exposed at the periplasmic surface. J. Bacteriol. **178**: 3156-3167.

12. **Fernandez, D., G. M. Spudich, X.-R. Zhou, and P. J. Christie.** 1996. The *Agrobacterium tumefaciens* VirB7 lipoprotein is required for stabilization of VirB proteins during assembly of the putative T-complex transport apparatus. J. Bacteriol. **178:** 3168-3176.

13. **Finberg, K. E., T. R. Muth, S. P. Young, H. B. Maken, S. M. Heitritter, A. N. Binns, and L. M. Banta.** 1995. Interactions of VirB9, -10, and -11 with the membrane fraction of *Agrobacterium tumefaciens*: Solubility studies provide evidence for tight associations. J. Bacteriol. **177:** 4881-4889.

14. **Fullner, K. J., K. M. Stephens, and E. W. Nester.** 1994. An essential virulence protein of *Agrobacterium tumefaciens*, VirB4, requires an intact mononucleotide binding domain to function in transfer of T-DNA. Mol. Gen. Genet. **245:** 704-715.

15. **Gennity, J. M., and M. Inouye.** 1991. The protein sequence responsible for lipoprotein localization in *Escherichia coli* exhibits remarkable specificity. J. Biol. Chem. **266:** 16458-16464.

16. **Giam, C. Z., S. Hayashi, and H. C. Wu.** 1984. Characterization of a novel lipoprotein mutant in *Escherichia coli*. J. Biol. Chem. **259**: 5601-5605.

17. **Haase, J., R. Lurz, A. M. Grahn, D. H. Bamford, and E. Lanka.** 1995. Bacterial conjugation mediated by plasmid RP4: RSF1010 mobilization, donor-specific phage production, and pilus production require the same Tra2 core components of a proposed DNA transport complex. J. Bacteriol. **177:** 4779-4791.

18 **Hatch, T. P.** Disulfide cross-linked envelope proteins: the functional equivalent of peptidoglycan in Chlamydiae? 1996. J. Bacteriol. **178:** 1-5.

19. **Hayashi, S., and H. C. Wu.** 1990. Lipoproteins in bacteria. J. Bioenergetics and Biomembranes. **22:** 451-471.

20. **Hooykaas, P. J. J., and A. G. M. Beijersbergen.** 1994. The virulence system of *Agrobacterium tumefaciens*. Annu Rev. Phytopathol. **32:** 157-179.

21. **Kado, C. I.** 1993. *Agrobacterium*-mediated transfer and stable incorporation of foreign genes in plants. p. 243-254. *In* D. B. Clewell (ed.), Bacterial Conjugation. Plenum Press, New York, N.Y.

22. **Kuldau, G. A., G. DeVos, J. Owen, G. McCaffrey, and P. Zambryski.** 1990. The *virB* operon of *Agrobacterium tumefaciens* pTiC58 encodes 11 open reading frames. Mol. Gen. Genet. **221:** 256-266.

23. **Lee, G. F., M. R. Lebert, A. A. Lilly, and G. L. Hazelbauer.** 1995. Identification of functionally important helical faces in transmembrane segments by scanning mutagenesis. Proc. Natl. Acad. Sci. USA **92:** 5416-5420.

24. **Lessl, M., and E. Lanka.** 1994. Common mechanisms in bacterial conjugation and Ti-mediated T-DNA transfer to plant cells. Cell **77:** 321-324.

25. **Lessl, M., D. Balzer, W. Pansegrau, and E. Lanka.** 1992. Sequence similarities between the RP4 Tra2 and the Ti VirB region strongly support the conjugation model for T-DNA transfer. J. Biol. Chem. **267:** 20471-20480.

26. **Matsuyama, S., T. Tajima, and H. Tokuda.** 1995. A novel periplasmic carrier protein involved in the sorting and transport of *Escherichia coli* lipoproteins destined for the outer membrane. EMBO J. **14:** 3365-3372.

27. **Peek, J. A. & R. K. Taylor.** 1992. Characterization of a periplasmic thiol:disulfide interchange protein required for the functional maturation of secreted virulence factors of *Vibrio cholerae.* Proc. Natl. Acad. Sci. USA **89:** 6210-6214.

28. **Pohlman, R. F., H. D. Genetti, and S. C. Winans. 1994.** Common ancestry between IncN conjugal transfer genes and macromolecular export systems of plant and animal pathogens. Mol. Microbiol. **14:** 655-668.

29. **Pugsley, A. P.** 1993. The complete general secretory pathway in gram-negative bacteria. Microbiol. Rev. **57**: 50-108.

30. **Rashkova, S., G. M. Spudich, and P. J. Christie.** Unpublished data.

31. **Riordon, J. F., and B. L. Vallee.** 1972. Reactions with *N*-Ethylmaleimide and *p*-Mercuribenzoate. *Methods Enzymol.* **25:** 449-456.

32. **Shirasu, K., and C. I. Kado.** 1993. Membrane location of the Ti plasmid VirB proteins involved in the biosynthesis of a pilin-like

conjugative structure on *Agrobacterium tumefaciens*. FEMS Micro. Lett. **111:** 287-294.

33. **Shirasu, K., P. Morel, and C. I. Kado.** 1990. Characterization of the *virB* operon of an *Agrobacterium tumefaciens* Ti plasmid: nucleotide sequence and protein analysis. Mol. Microbiol. **4:** 1153-1163.

34. **Spudich, G. M., D. Fernandez, X.-R. Zhou, and P. J. Christie.** Definition of two steps in the biogenesis of the *Agrobacterium tumefaciens* T-complex transport apparatus: Stabilization of VirB7 homodimers and of VirB7/VirB9 heterodimers by intermolecular disulfide bonds. Proc. Natl. Acad. Sci. USA. Submitted.

35. **Stachel, S. E., and E. W. Nester.** 1986. The genetic and transcriptional organization of the *vir* region of the A6 Ti-plasmid of *Agrobacterium tumefaciens*. EMBO J. **5:** 1445-1454.

36. **Stephens, K. M., C. Roush, and E. Nester.** 1995. *Agrobacterium tumefaciens* VirB11 protein requires a consensus nucleotide binding site for function in virulence. J. Bacteriol. **177:** 27-36.

37. **Thompson, D. V., L. S. Melchers, K. B. Idler, R. A. Schilperoort, and P. J. J. Hooykaas.** 1988. Analysis of the complete nucleotide sequence of the *Agrobacterium tumefaciens virB* operon. Nucleic Acids Res. **16:** 4621-4636.

38. **Thorstenson, Y. R., G. A. Kuldau, and P. C. Zambryski.** 1993. Subcellular localization of seven VirB proteins of *Agrobacterium tumefaciens*: Implications for the formation of a T-DNA transport structure. J. Bacteriol. **175**: 5233-5241.

39. **Ward, J. E., D. E. Akiyoshi, D. Regier, A. Datta, M. P. Gordon, and E. W. Nester.** 1988. Characterization of the *virB* operon from an *Agrobacterium tumefaciens* Ti plasmid. J. Biol. Chem. **263:** 5804-5814.

40. **Ward, J. E., D. E. Akiyoshi, D. Regier, A. Datta, M. P. Gordon, and E. W. Nester.** 1990. Characterization of the *virB* operon from an *Agrobacterium tumefaciens* Ti plasmid. J. Biol. Chem. **265:** 4768.

41. **Ward, J. E., Jr., E. M. Dale, E. W. Nester, and A. N. Binns.** 1990. Identification of a VirB10 protein aggregate in the inner

membrane of *Agrobacterium tumefaciens.* J. Bacteriol. **172:** 5200-5210..

42. **Weiss, A. A., Johnson, F. D., and D. L. Burns.** 1993. Molecular characterization of an operon required for pertussis toxin secretion. Proc. Natl. Acad. Sci. USA **90**: 297-2974.

43. **Winans, S. C., D. L. Burns, and P. J. Christie.** Adaptation by bacterial pathogens of a conjugal transfer system for export of macromolecules. Trends Microbiol. In press.

44. **Zupan, J. R., and P. Zambryski.** 1995. Transfer of T-DNA from *Agrobacterium* to the plant cell. Plant Physiol. **107:** 1041-1047.

CHAPTER 7

Agrobacterium and Plant Genes Affecting T-DNA Transfer and Integration

Soma B. Narasimhulu, Jaesung Nam, Xiao-bing Deng, Rodrigo Sarria, Walt Ream, and Stanton B. Gelvin

Department of Biological Sciences, Purdue University
West Lafayette, Indiana 47907-1392

During the inception of crown gall tumorigenesis, *Agrobacterium tumefaciens* processes a region of DNA (the transferred, or T-DNA) from the resident tumor-inducing (Ti) plasmid and transfers this DNA to plant cells. Proteins encoded by the virulence (*vir*) region of the Ti plasmid regulate T-DNA processing and transfer. Nicking of 25-bp directly repeated T-DNA "border" sequences by the VirD2 endonuclease results in the generation of single-stranded T-DNA molecules (T-strands) with which VirD2 is tightly associated at the 5' end (11, 14, 37, 40). These single-stranded DNA molecules are transferred to the plant cytoplasm (41), perhaps as a complex (the T-complex; 13) with the single-stranded DNA binding protein VirE2 (2, 3, 6, 10). Targeting of the T-DNA to the plant nucleus may be mediated by nuclear localization sequences (NLS) within the associated VirD2 and VirE2 proteins (5, 12, 15, 18, 27, 28, 34). T-DNA molecules eventually integrate into the plant chromosomes, thereby stabilizing the oncogenes and opine biosynthesis genes encoded by the T-DNA. However, non-integrated copies of T-DNA may persist in the nucleus for a period of time.

Although we now have a fairly detailed understanding of the early events that occur in *A. tumefaciens* resulting in *vir* gene induction, T-DNA processing, and T-DNA, little is known about the events that take place within the plant cell involving the targeting of T-DNA to the nucleus, its ultimate integration into plant nuclear DNA, and the early stages of T-DNA expression. Similarly, our knowledge of the role that plant genes play in these processes is scant. In an attempt to elucidate the kinetics of T-DNA transport to the nucleus as well as to determine which Vir proteins are necessary to effect this process, we developed a very sensitive reverse transcriptase-PCR assay to detect T-DNA transcription early after infection. Using this assay, we detected T-DNA transcription 18 hr after cocultivation of tobacco suspension culture cells with *A. tumefaciens*. Transient transcription of the T-DNA absolutely depends upon intact VirB, VirD1/VirD2, and VirD4 proteins. Transient transcription decreased without intact VirC and VirE2 proteins. A mutation of VirD2 that eliminates the ω domain (28) resulted in an *A. tumefaciens* strain that has very weak virulence, but we show that it still transiently directs a high level of *gusA* transcription. These results suggest that the ω domain of VirD2 is at least partially responsible for the stabilization of T-DNA transcription, perhaps by aiding T-DNA integration. The initial kinetics of *gusA* transcription in maize suspension culture cells are similar to those of tobacco cells, suggesting that T-DNA transfer, nuclear targeting, and replication to a double-stranded form occur in maize in a manner similar to that of tobacco. In maize, however, transcription of the *gusA* gene is highly transient, suggesting that the block to stable transformation of maize by *A. tumefaciens* occurs at the stage of T-DNA integration. Finally, we show that variation in susceptibility to tumorigenesis among *Arabidopsis thaliana* ecotypes is a heritable trait. One recalcitrant ecotype is deficient in the ability to integrate T-DNA into the plant genome.

Development of a system to detect the early transcription of T-DNA

To determine the kinetics of transport of the T-DNA from the cytoplasm to the nucleus and the importance of various Vir proteins in this process, we developed a sensitive assay to detect early transcription of the T-DNA. We constructed a T-DNA binary vector, pBISN1 (Figure 1), that contains a *gusA* (*uidA*) gene under the transcriptional control of a super-promoter (24). The *gusA* gene contains a 189 bp intron from the potato ST-LS1 gene (35). We generated polymerase chain reaction primers that would amplify a 732 bp region containing this intron. However, if the intron were processed from a *gusA* transcript, we would amplify (using reverse transcriptase-PCR [RT-PCR]) a 543 bp fragment. Because of the size difference in amplification products from the *gusA*-intron gene and the processed *gusA* transcript, we could distinguish between *gusA*

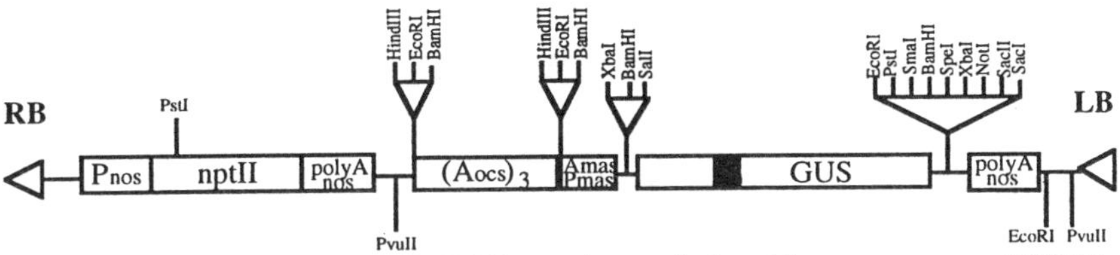

Figure 1. Map of the T-DNA region of the binary vector pBISN1. pBISN1 is based upon the plasmid pBI101.2. RB, T-DNA right border; LB, T-DNA left border; p_{nos}, nopaline synthase promoter; nptII, neomycin phosphotransferase II coding region; $polyA_{nos}$, nopaline synthase polyadenylation signal sequence; $(A_{ocs})3$, trimer of the octopine synthase activating element; A_{mas}, mannopine synthase activating element; P_{mas}, mannopine synthase promoter; GUS, β-glucuronidase coding region. The filled region within the GUS gene represents the ST-LS1 intron.

mRNA and *gusA*-intron DNA from contaminating *A. tumefaciens* cells. RT-PCR analysis of in vitro-transcribed *gusA* RNA indicated that, by using this assay, we could readily detect 10 fg of *gusA* mRNA (data not shown).

We infected rapidly dividing tobacco BY-2 suspension culture cells (approximately 20 bacterial cells per tobacco cell) with each of three *A. tumefaciens* strains harboring pBISN1. *A. tumefaciens* At793 lacks a Ti plasmid, *A. tumefaciens* At789 contains the octopine-type Ti plasmid pTiA6, and *A. tumefaciens* At790 contains the agropine-type supervirulent Ti plasmid pTiBo542. After 12 hr of cocultivation, we pelleted the plant cells by centrifugation, washed them in plant culture medium, and resuspended them in plant culture medium containing antibiotics to kill any remaining bacterial cells. We isolated total cellular RNA from infected plant cells at various times starting from the initiation of cocultivation, and subjected the RNA to RT-PCR using primers that would amplify the region of *gusA* mRNA flanking the (processed) intron.

Figure 2 (A and B) shows that we could detect *gusA* transcripts beginning 20-24 hr after infection of tobacco BY-2 suspension culture cells by either *A. tumefaciens* strain containing a Ti plasmid. The level of *gusA* mRNA increased and peaked at 36 hr, after which there was a slight decline in *gusA* mRNA. We could not detect *gusA* mRNA in noninfected plant cells and in plant cells from which RNA was extracted immediately after the start of cocultivation (0 hr time point). The synthesis of *gusA* mRNA in cocultivated plant cells depended upon the presence of a Ti plasmid within the infecting *A.*

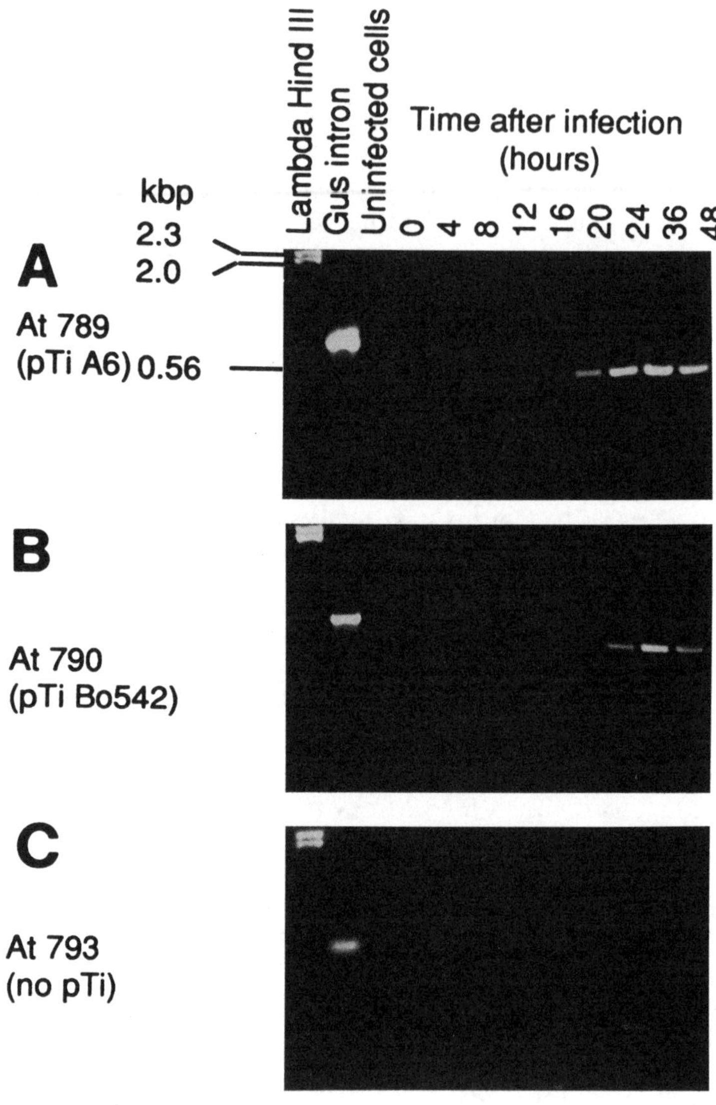

Figure 2. Kinetics of *gusA* gene expression (until 48 hr) in tobacco suspension culture cells. Tobacco cells were infected with *A. tumefaciens* At789 (panel A), At790 (panel B), or At793 (panel C) for various periods of time. RNA was extracted, subjected to RT-PCR, and analyzed by agarose gel electrophoresis. GUS intron, PCR amplification of a 732 bp *gusA*-intron gene region.

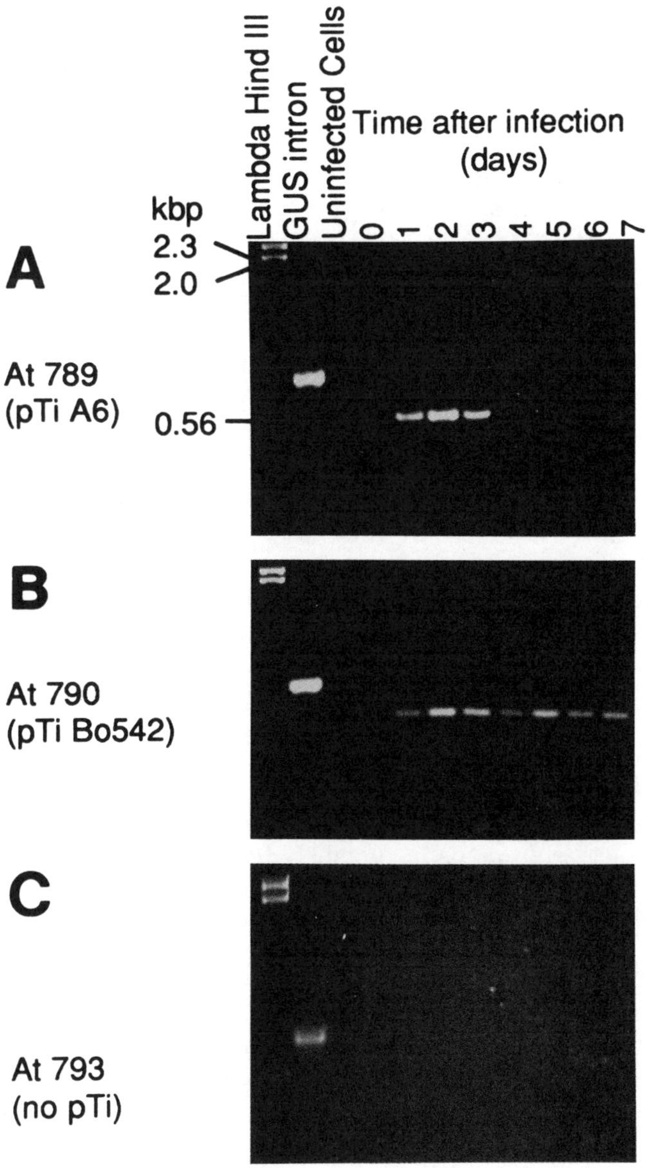

Figure 3. Kinetics of *gusA* gene expression (until 7 days) in tobacco suspension culture cells. Tobacco cells were infected with *A. tumefaciens* At789 (panel A), At790 (panel B), or At793 (panel C) for 24 hr, the cells were washed free of bacteria, and were resuspended in culture medium plus antibiotics for various periods of time. RNA was extracted and subjected to RT-PCR.

tumefaciens strain. Repeated analyses of RNA extracted from tobacco cells cocultivated with *A. tumefaciens* At793 (lacking a Ti plasmid) never revealed the presence of *gusA* mRNA (Figure 2C). These data suggest that *gusA* gene expression was *vir* gene-dependent.

We next determined the stability of *gusA* gene expression after infection with *A. tumefaciens*. We cocultivated tobacco BY-2 cells with various *A. tumefaciens* strains, washed the cells 24 hr after infection, resuspended the cells in culture medium containing antibiotics to kill the bacteria, and isolated total cellular RNA at 24-hour intervals starting from the initiation of cocultivation. Figure 3A shows that, using *A. tumefaciens* At789 (containing pTiA6), we could detect *gusA* mRNA for up to 3 days following the initiation of cocultivation. We could never detect *gusA* transcripts after 72 hr of infection. Using *A. tumefaciens* At790 (containing pTiBo542), however, we routinely detected *gusA* mRNA for at least 7 days (the latest time point that we assayed; Figure 3B).

Dependence of early T-DNA transcription upon *vir* gene function

To determine the importance of different virulence proteins in the early stages of T-DNA transfer, transcription, and perhaps integration, we introduced pBISN1 into *A. tumefaciens* containing mutations in various *vir* genes. We incubated each mutant strain separately with tobacco BY-2 suspension culture cells, washed the plant cells and killed the bacteria with antibiotics after 12 hr cocultivation, extracted tobacco RNA 18, 24, 36, and 48 hr after the initiation of cocultivation, and assayed for the presence of *gusA* mRNA by RT-PCR.

Figure 4B shows that *A. tumefaciens* At805 mutant in *virB* could not direct the synthesis of *gusA* mRNA in infected tobacco cells. We have previously shown that this same strain could not transfer T-DNA to the cytoplasm of tobacco cells (41), although wild-type levels of T-strands accumulated in protoplast-induced (36) or acetosyringone-induced (31) bacterial cells. Similarly, *A. tumefaciens* At819, containing a mutant *virD4* gene, could not effect *gusA* transcription in cocultivated tobacco cells (Figure 4H). VirD4 is a bacterial periplasmic membrane-localized protein (26). *A. tumefaciens* or *E. coli* strains containing mutant *virD4* genes can process T-DNA and accumulate T-strands (17, 31). *A. tumefaciens* cells mutant in *virD4* may not, however, be able to export T-strands through the VirB protein export apparatus. *A. tumefaciens* At806, containing a polar insertion in *virD1*, could not direct *gusA* mRNA synthesis (Figure 4E). An *A. tumefaciens* strain containing this same *virD1* mutation could not process T-DNA or accumulate T-strands (31, 36).

Cocultivation of tobacco cells with *A. tumefaciens* At807 (mutant in *virE*) resulted in only a low level of *gusA* mRNA accumulation late (48 hr) in infection (Figure 4D). Stachel and Nester (29) have previously shown that similar mutations in *virE* render the bacterium severely attenuated in virulence, despite the fact that normal levels of T-strands accumulate in induced cells (31, 36). Thus, the presence of VirE2 single-stranded DNA binding protein is not necessary to protect the T-strand within the bacterium. The low level of *gusA* transcripts that accumulate late after infection reflect the low steady-state level of single-stranded T-DNA that accumulates in the cytoplasm of infected tobacco cells (41).

Mutations in *virC* severely attenuate the virulence of *A. tumefaciens* on most plant species, including tobacco (Stachel and Nester, 1986), although nearly wild-type levels of T-strands accumulate in induced bacterial cells (31, 36). *A. tumefaciens* At871, mutant in *virC*, could direct only a very low level of transient *gusA* mRNA accumulation in infected tobacco cells (Figure 4C).

A. tumefaciens containing the non-polar insertion mx304 in the 3' end of *virD2* are avirulent (18, 29), although they accumulate T- strands following induction by acetosyringone (29). The Tn*3*-HoHo1 insertion occurs 73 amino acids from the C-terminus of VirD2 and translationally fuses β-galactosidase protein to the C-terminus of VirD2 (18). The resulting altered VirD2 protein lacks the C-terminal NLS (15, 34) and the ω region that is important for efficient tumorigenesis in potato (28) and tobacco (S. Gelvin, unpublished data). Therefore, we were surprised to detect relatively high levels of *gusA* transcripts in tobacco cells infected with *A. tumefaciens* At808 harboring this *virD2* mutation (Figure 4F). Quantitative RT-PCR indicated that the level of *gusA* transcript resulting from infection of tobacco cells by this mutant *A. tumefaciens* strain was 20-30% that found in tobacco cells cocultivated by *A. tumefaciens* At789 containing a wild-type *virD2* gene (data not shown). The

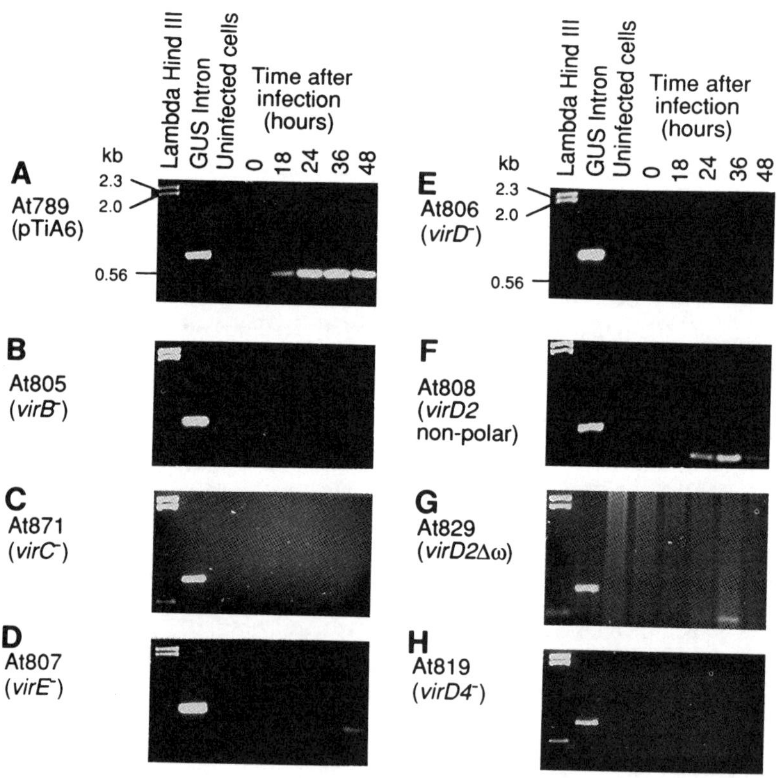

Figure 4. Effect of *vir* gene mutations upon the kinetics of *gusA* gene expression (until 48 hr) in tobacco suspension culture cells. Tobacco cells were infected with *A. tumefaciens* At789 (panel A), At805 (panel B), At871 (panel C), At807 (panel D), At806 (panel E), At808 (panel F), At829 (panel G), or At819 (panel H) for various periods of time. RNA was extracted, subjected to RT-PCR, and analyzed by agarose gel electrophoresis.

presence of a high level of *gusA* mRNA in these cells was considerably more transient than that resulting from infection by the wild-type bacterium. These data indicate that deletion of the carboxy-terminal NLS and/or ω region of VirD2 results in high level transient expression of T-DNA-encoded genes. The high level of transient *gusA* gene expression following infection of tobacco cells by *A. tumefaciens* At808 suggests that, despite the lack of a carboxy-terminal NLS in the VirD2 protein encoded by this strain, the T-DNA is efficiently directed to the plant nucleus.

The translational fusion of β-galactosidase protein to the C-terminus of VirD2 in *A. tumefaciens* At808 could result in a protein with altered properties. Therefore, we repeated the tobacco cell infections using *A. tumefaciens* At829. This strain encodes a mutant VirD2 protein containing two serine residues in place of four of the five amino acids in the ω region (28); therefore, it lacks the ω region while still retaining the NLS. Infection of potato tuber discs (28) or tobacco leaf sections (S. Gelvin, unpublished) with a similar *A. tumefaciens* strain harboring this same *virD2* mutation resulted in 3.3% and 2.5%, respectively, the number of tumors compared to infection by a strain harboring a wild-type *virD2* gene. Figure 4G shows that *A. tumefaciens* At829 directed a high level of transient transcription of the *gusA* gene. Quantitative RT-PCR indicated that, as with *A. tumefaciens* At808, *gusA* transcripts accumulated to 20-30% the level as when we infected tobacco cells with the wild-type strain *A. tumefaciens* At822 (data not shown). These data indicate that the ω region of VirD2 is not required for T-DNA transport to the plant cytoplasm or nucleus, replication to a double-stranded form, or transcription. The data suggest, however, that this region of VirD2 may be involved in the stabilization of T-DNA transcription, perhaps by mediating T-DNA integration into the plant genome.

Transient expression of *gusA* mRNA in maize suspension culture cells

We cocultivated maize Black Mexican Sweet (BMS) suspension culture cells with *A. tumefaciens* At789 (harboring pTiA6) or At790 (harboring pTiBo542) that had been previously induced with acetosyringone. Figure 5 shows that, using RT-PCR, we could detect *gusA* transcripts using either *A.*

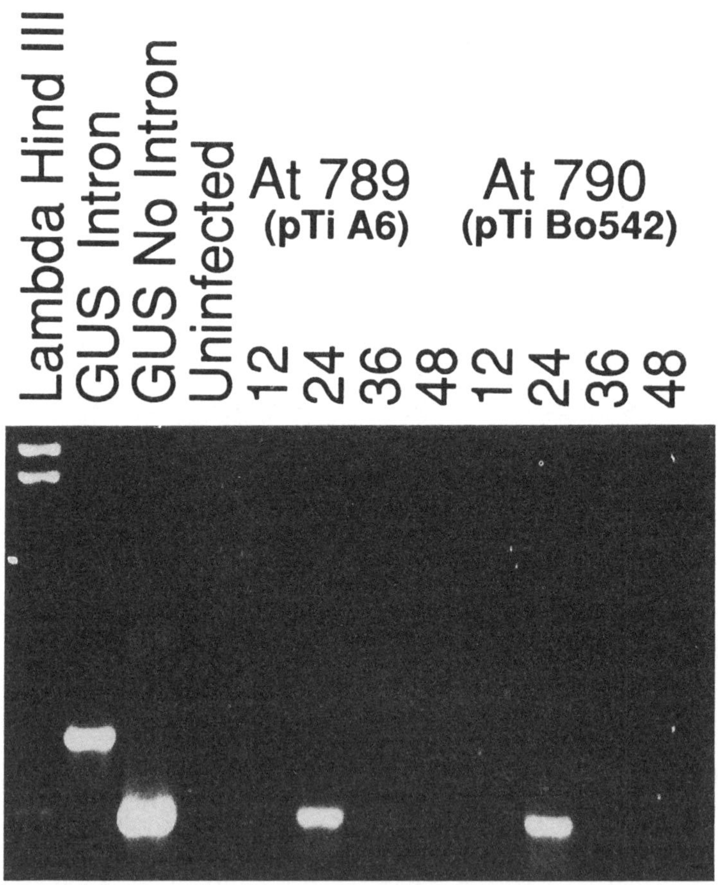

Figure 5. Kinetics of *gusA* gene expression (until 48 hr) in maize suspension culture cells. Maize cells were infected with *A. tumefaciens* At789 or At790 for various periods of time. RNA was extracted, subjected to RT-PCR, and analyzed by agarose gel electrophoresis.

tumefaciens strain within infected maize cells 24 hr after infection. The appearance of *gusA* mRNA was, however, very transient. By 36 hr after the initiation of cocultivation, *gusA* mRNA greatly decreased and was never detected at 48 hr. We detected the same kinetics of transcript accumulation and disappearance in three independent cocultivation experiments. Thus, the kinetics of initial appearance of *gusA* mRNA in maize approximated that of tobacco. In maize, however, the presence of *gusA* mRNA was highly transient, suggesting that in this monocot species T-DNA integration and/or stable transcription was defective.

Arabidopsis thaliana ecotype variation and susceptibility to crown gall tumorigenesis

As an important vector system for plant genetic transformation, scientists have intensively studied for several decades the mechanisms by which *A. tumefaciens* transfers T-DNA to the plant cell and induces neoplastic cell growth. By using a large number of *Agrobacterium* strains and mutants, we now understand reasonably well many of the early events of crown gall tumorigenesis, including the induction of the *vir* (virulence) genes (39), processing of the T-DNA from the Ti plasmid (8, 30), and the formation of bacterial channels for exporting the T-DNA (19, 32, 38), possibly as a DNA-protein complex (the T-complex; 13). In contrast, little is known about plant host factors involved in crown gall tumorigenesis. There are at least three obstacles that *A. tumefaciens* must overcome to transform a plant cell. First of all, *A. tumefaciens* must transfer T-DNA into the cytoplasm of plant cells after the DNA has crossed the plant cell wall and plasma membrane. We now know that T-DNA enters the plant as a single-stranded DNA molecule (33, 41). Before genes encoded by the T-DNA can be expressed, the T-DNA must reach the plant nucleus. These events may be aided by nuclear localization signals (NLS) found in VirD2 and VirE2 proteins that may accompany the T-DNA into the plant cell (4, 5, 12, 15, 18, 27, 28, 34). Finally, the T-DNA must become stabilized by integration into the plant genome. T-DNA integration may occur by illegitimate recombination (9, 21, 22, 25). Should any of these steps in the infection cascade fail, the result would be an abortive infection. We currently know little, however, regarding whether or what kind of specific plant factors in the cytosol and/or nucleus are involved in these processes.

Recently, a number of laboratories have used different ecotypes of *Arabidopsis thaliana* as a model plant system to investigate differential host response to various strains or races of bacteria, fungi, and viruses. We wished to determine whether differences also exist among *Arabidopsis* ecotypes with regard to tumorigenesis by *A. tumefaciens*, and what the biochemical and genetic basis of such differences were. We therefore developed an *in vitro* root inoculation assay and screened *Arabidopsis thaliana* ecotypes for susceptibility or resistance to crown gall disease. We have identified several ecotypes that are hyper-susceptible to crown gall tumorigenesis, as well as other ecotypes that are recalcitrant (less susceptible) to tumorigenesis. Interestingly, we have identified several ecotypes that appear to permit T-DNA transfer and nuclear transport, but are deficient in T-DNA integration.

Response of *Arabidopsis thaliana* ecotypes to different *Agrobacterium tumefaciens* strains

We initially characterized the response of 11 *Arabidopsis* ecotypes to four different *A. tumefaciens* strains using an *in vitro* root segment infection tumorigenesis assay. These ecotypes were Aa-0, Be-0, Col-0, C-24, La-er, Lip, No-0, Oy-0, RLD, an unidentified ecotype, and "WS". Subsequent analysis indicated that "WS" was misclassified, and is not the ecotype Wassilewskija. We thus denote this unknown ecotype "WS" rather than WS.

We tested four *A. tumefaciens* strains. All strains had the same C58 chromosomal background. *A. tumefaciens* A136 lacks a Ti plasmid and is therefore avirulent, whereas the virulent strains *A. tumefaciens* A348, A208, and A281 contain the octopine-type Ti plasmid pTiA6, the nopaline-type Ti plasmid pTiT37, or the agropine-type supervirulent Ti plasmid pTiBo542, respectively. We chose *A. tumefaciens* A208 as the best strain for this study because it caused more severe symptom development in most of the ecotypes we tested than did the other bacterial strains. In addition, tumors induced by A208 tended to develop into teratomas, with clusters of modified leaves appearing over the entire surface of the calli. This teratoma phenotype was useful to distinguish actual tumors from non-

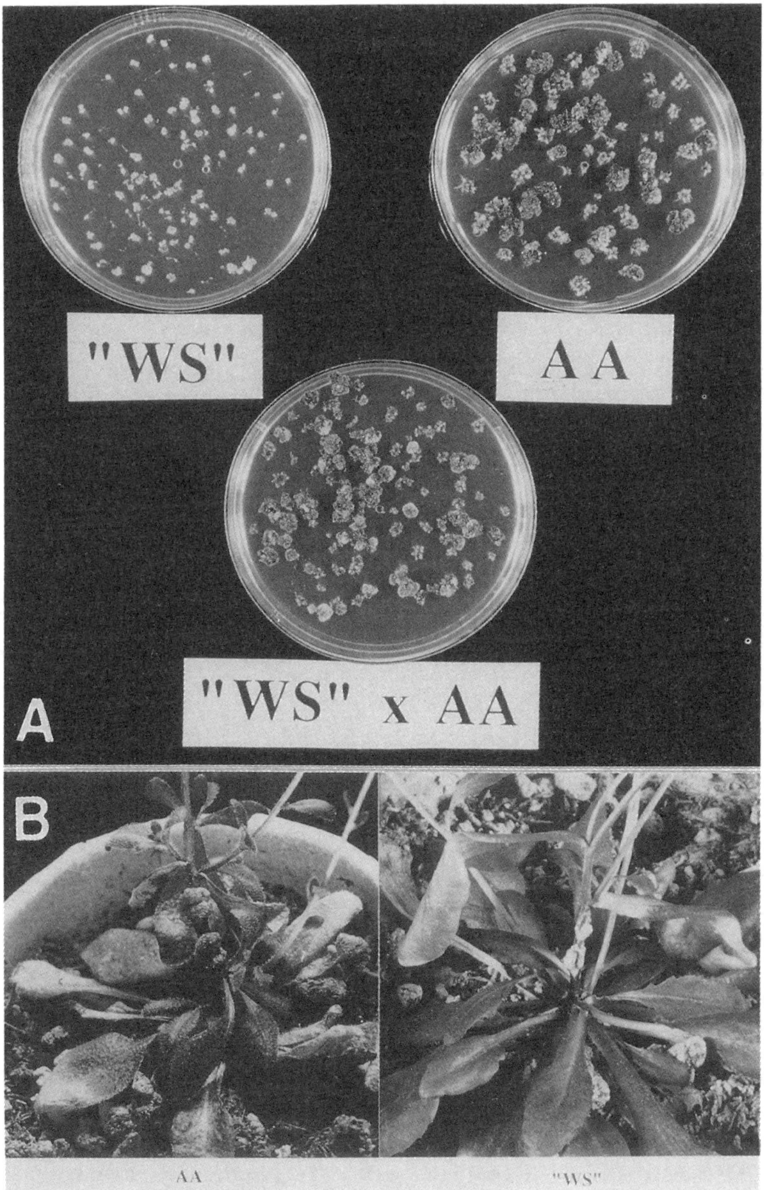

Figure 6. Tumorigenesis of *A. tumefaciens* upon *Arabidopsis thaliana* root segments (**A**) and flower bolts (**B**). Sterile root segments or flower bolts of ecotypes Aa-0, "WS", or the F1 progeny of a cross between these two ecotypes were inoculated with *A. tumefaciens* A208 and the resulting tumors photographed one month later.

transformed calli that sometimes grew even in response to the avirulent strain *A. tumefaciens* A136.

From among these 11 *Arabidopsis* ecotypes, "WS" appeared recalcitrant to tumorigenesis by *A. tumefaciens* A208. Usually, infected root segments of ecotype "WS" did not respond to inoculation by this bacterial strain. Occasionally, a few small yellow tumorous calli lacking teratomas developed.In contrast, infected root segments of ecotype Aa-0 consistently developed large green tumors with teratomas using this assay (Figure 6A). In addition, ecotype "WS" exhibited a greatly reduced tumorigenesis response, relative to ecotype Aa-0, when we inoculated bacteria onto a wounded flower stalk (Figure 6B). We therefore concentrated our initial analyses upon these two ecotypes.

Comparison of DNA and protein synthesis rates between *Arabidopsis* ecotypes Aa-0 and "WS"

Because plant growth rate at the time of inoculation is an important factor in determining gall formation in *Pinus radiata* and because plant cell division is required for transformation (1), we compared the rates of DNA and protein synthesis between the ecotypes Aa-0 and "WS". The incorporation of ^{3}H-thymidine or ^{14}C-amino acids was approximately linear for at least 5 hours. One day pre-incubation of the roots on CIM prior to incubation with isotope did not result in a major effect upon ^{3}H-thymidine incorporation. Preincubation of the roots on CIM resulted in a 70% increase in the extent of incorporation of ^{14}C-amino acids in the ecotype Aa-0 but not in "WS", although the eventual rates of incorporation were approximately equal. Nonetheless, we did not find a major difference in the incorporation of these isotopes between the ecotypes Aa-0 and "WS" (data not shown). We therefore conclude that the difference in tumorigenesis between these ecotypes does not result from a major difference in cellular metabolism.

Induction of *A. tumefaciens vir* genes by compounds secreted by the roots of various *Arabidopsis* ecotypes

We investigated the ability of various *Arabidopsis* ecotypes to induce *vir* genes. We introduced a *virH::lacZ* fusion (29) into *A. tumefaciens* and monitored β-galactosidase activity following co-cultivation of *A. tumefaciens* At44 with *Arabidopsis* root segments. There was no substantial difference in *vir* gene induction among the ecotypes tested. In addition, pre-incubation of roots on CIM for 1 or 3 days prior to co-cultivation with *A. tumefaciens* did not have a substantial effect upon *vir* gene induction (data not shown). Thus, we could not attribute the difference in tumorigenesis between Aa-0 and "WS" to a difference in the ability of these ecotypes to induce *A. tumefaciens vir* genes.

Sensitivity of *Arabidopsis* root growth to phytohormones

Because the growth of *Arabidopsis* roots is extremely sensitive to phytohormones and tumorigenesis occurs as a response of plant cells to these hormones, we compared the hormone sensitivity of Aa-0 and "WS" by measuring the relative root growth of seedlings of these ecotypes grown vertically on plates containing different concentrations of phytohormones.

The inhibition of primary root growth of the ecotypes Aa-0 and "WS" was similar at several different concentrations of kinetin, 2,4-D, and IAA, respectively (data not shown). We also investigated primary root growth on various dilutions of CIM. CIM contains kinetin (1.4 μM), 2,4-D (2.3 μM), and IAA (28.5 μM), and growth of roots on 1x CIM results in callus formation rather than root elongation. Pre-incubation of roots on CIM prior to infection by *A. tumefaciens* stimulates tumor induction using the *in vitro* root segment inoculation test. Various dilutions of the hormone combination found in CIM inhibit root growth of both ecotypes in a similar manner (data not shown). Thus, it is not likely that the difference in tumorigenesis between Aa-0 and "WS" results from a differential sensitivity of the cells of these ecotypes to the phytohormones auxin and cytokinin.

Efficiency of T-DNA transfer to cells of the *Arabidopsis* ecotypes Aa-0 and "WS"

To investigate whether the difference in tumorigenesis between *Arabidopsis* ecotypes Aa-0 and "WS" results from a deficiency in T-DNA transfer from *A. tumefaciens* to "WS" cells, we investigated the transient T-DNA-mediated transfer to and expression of a *gusA* gene in the cells of these two ecotypes. We characterized β-glucuronidase (GUS) expression in *Arabidopsis* cells using both a qualitative histochemical staining assay and a quantitative fluorimetric assay. We introduced into *A. tumefaciens* A208 the binary T-DNA vector pCNL65 (20). This vector contains a *gusA* gene, harboring an intron, under control of the CaMV 35S promoter. The intron in the *gusA* gene permits expression of GUS activity only in plant cells, but not in the bacteria (20). Root explants of the ecotypes Aa-0 and "WS" showed approximately equal numbers and intensity of blue spots following infection by *A. tumefaciens* A208(pCNL65) for 7 days (two days of co-cultivation, followed by five further days of incubation in the presence of antibiotics to kill the bacteria) and histochemical staining with X-gluc. Stained cells were preferentially localized at the cut termini of root explants because these cells were preferentially activated by wounding and accessible for T-DNA transfer (data not shown). We additionally quantitated GUS activity in these infected roots, and in the roots of other ecotypes, using a fluorimetric MUG assay. Figure 7 shows that ecotype "WS" actually displayed twice the level of GUS expression as did ecotype Aa-0 using this assay. GUS activity detected early after infection most probably represents transient expression of T-DNA that is not yet integrated into the plant genome (16, 20, 23). These data indicate that the efficiency of T-DNA transfer to and expression in the plant nucleus are similar between the ecotypes Aa-0 and "WS". We therefore conclude that the difference in tumorigenesis between these ecotypes does not result from a deficiency of T-DNA transfer or nuclear targeting in the ecotype "WS".

Stable GUS expression and kanamycin resistance in infected Aa-0 and "WS" cells

The finding that ecotypes Aa-0 and "WS" showed an approximately equal amount of transient GUS expression, but exhibited a different tumorigenesis response to *A. tumefaciens* infection, suggested to us that the ecotype "WS" may be deficient in some step of tumorigenesis that involves the stabilization of T-DNA and/or its expression in infected plant cells. To test this hypothesis, we examined two additional transformation events that require stable integration and expression of T-DNA in the plant cells: Stable GUS expression

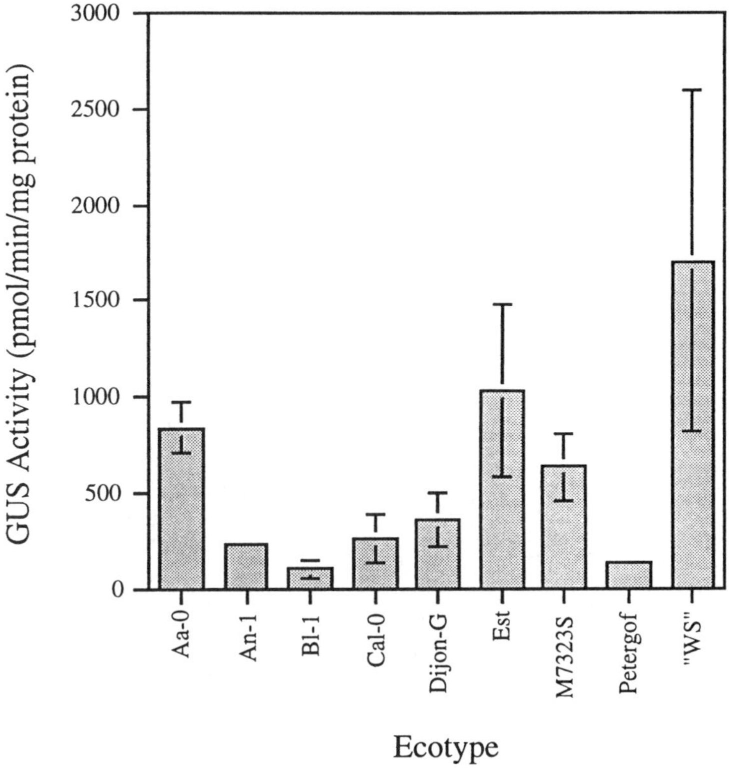

Figure 7. Transient GUS activity in the roots of various *Arabidopsis* ecotypes. *A. tumefaciens* A208 (pCNL65) was used to infect sterile root segments, the bacteria were killed with antibiotics after 2 days, and GUS activity measured 5 days later.

and kanamycin-resistant growth of calli derived from infected *Arabidopsis* root segments.

We co-cultivated root segments of the ecotypes Aa-0 and "WS" for two days with A208(pCNL65), then transferred the root segments to CIM to induce calli without selection or to CIMcontaining kanamycin to select stable transformants. Figure 8 shows that root segments only from ecotype Aa-0 developed calli on CIM containing kanamycin, although root segments of both ecotypes generated calli on CIM lacking kanamycin. We additionally investigated stable GUS expression by staining calli derived from root segments, grown on CIM without kanamycin selection, with X-gluc. Calli from ecotype "WS" showed only a few small blue spots. However, most of the calli from ecotype

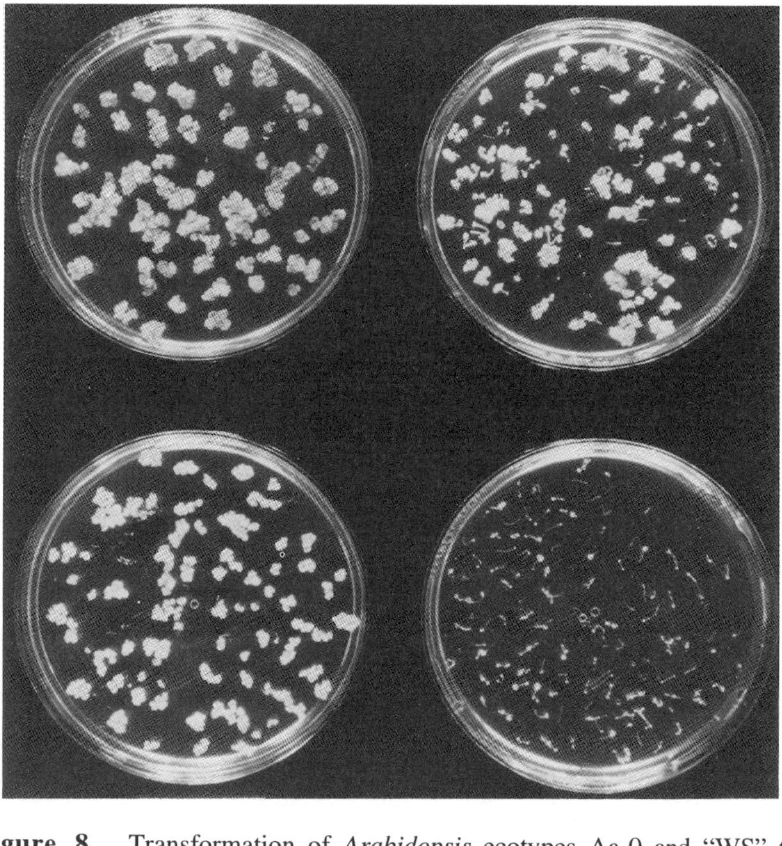

Figure 8. Transformation of *Arabidopsis* ecotypes Aa-0 and "WS" to kanamycin-resistance. Sterile root segments were inoculated with *A. tumefaciens* A208(pCNL65), the bacteria killed with antibiotics after 2 days cocultivation, and the root segments incubated on CIM containing kanamycin. The kanamycin-resistant calli were photographed after 1 month. Top row: ecotype Aa-0; Bottom row: ecotype "WS". Left: CIM lacking kanamycin; Right: CIM containing kanamycin.

Aa-0 displayed large patches of deep blue spots (data not shown).

Efficiency of T-DNA integration into the genomes of ecotypes Aa-0 and "WS"

The results presented above suggested that the basis for the recalcitrance of "WS" to tumorigenesis could be a deficiency of T-DNA integration into the plant genome. To investigate this possibility, we infected root segments of ecotypes Aa-0 and "WS" with A208(pCNL65) and grew calli on CIM without selection. These calli were used to initiate cell suspensions grown in liquid CIM, which were composed of a mixture of transformed and non-transformed cells. We isolated DNA from these cells, separated high molecular weight plant DNA from any possible contaminating pCNL65 binary vector DNA by electrophoresis though a 0.6% agarose gel, and blotted the DNA onto a nylon membrane. Figure 9A shows that, when first hybridized with a *gusA* gene probe, high molecular weight DNA from ecotype Aa-0 showed a much stronger signal than did DNA from ecotype "WS". Densitometric analysis of the autoradiogram indicated that this difference was approximately 5-fold. This result demonstrates that T-DNA could integrate into the DNA of ecotypes Aa-0 and "WS" to different extents.

Even though we killed *A. tumefaciens* cells with a combination of antibiotics (timentin, carbenicillin, vancomycin, and cefotaxime) before DNA isolation, we felt it important to show that the different signal intensities did not result from bacterial pCNL65 DNA contamination of the isolated *Arabidopsis* DNA. We therefore washed the *gusA* gene probe from the membrane and re-hybridized the membrane with a 2.3 kbp *Not*I-*Bgl*II DNA fragment located outside the left T-DNA border. We did not detect any hybridization signal in any *Arabidopsis* DNA lane (Figure 9B). This result indicates that the hybridization that we observed using the *gusA* gene probe did not result from contaminating bacterial DNA. Finally, we washed the 2.3 kbp hybridization probe from the membrane and re-hybridized the membrane with a PAL (phenylalanine-ammonia lyase) gene probe in order to show that we had transferred equal amounts of *Arabidopsis* DNA to each lane of the membrane. Figure 9C shows that, indeed, each lane contained an approximately equal amount of DNA.

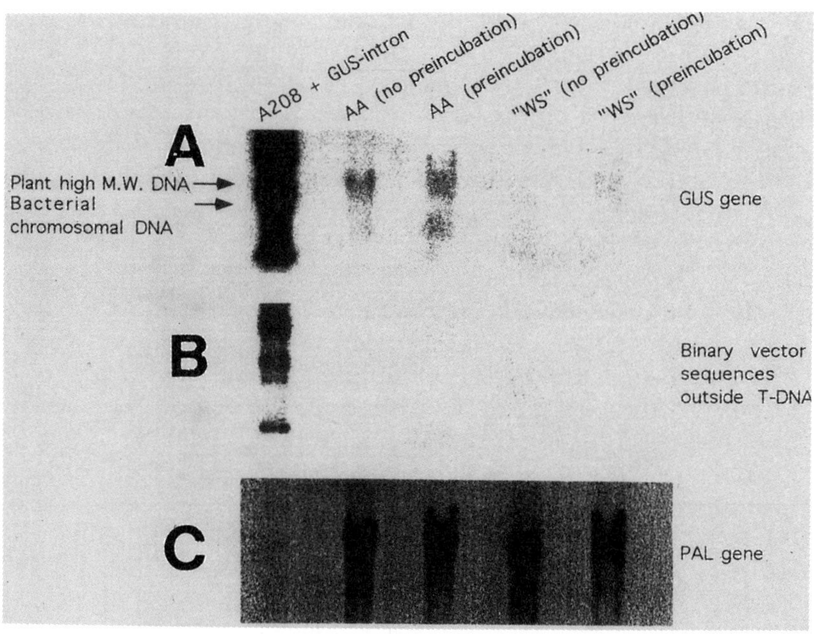

Figure 9. Integration of T-DNA into high molecular weight *Arabidopsis* DNA. Root segments of *Arabidopsis* ecotypes Aa-0 and "WS" were infected with *A. tumefaciens* A208(pCNL65), the bacteria were killed with antibiotics after 2 days cocultivation, and calli grown on CIM under non-selective conditions. High molecular weight plant DNA was isolated, subjected to electrophoresis, blotted onto nitrocellulose, and hybridized with various probes. (A) The probe is from the *gusA* gene; (B) The probe is from the T-DNA binary vector backbone; (C), The probe is an *Arabidopsis* PAL gene.

Sensitivity of ecotypes Aa-0 and "WS" to ionizing radiation

DNA damaging agents, especially ionizing radiation, can cause double-strand or single-strand breaks in plant chromosomes. Chromosome breaks can be repaired imprecisely in plants through a non-homologous recombination mechanism. Because T-DNA is suspected to be integrated into the plant genome through an illegitimate recombination process (9, 21, 22, 25), we investigated the sensitivity of ecotypes Aa-0 and "WS" to γ-radiation. Our rationale for this experiment was that this sensitivity may reflect the ability of the plant to integrate T-DNA into the plant genome.

We exposed five day-old seedlings of Aa-0 and "WS", which had only two expanded cotyledons, to different amounts of γ-radiation, following which we incubated the plants for an additional two weeks. We measured the sensitivity of the plants to severe radiation damage by their ability to produce the first two rosette leaves. Radiation treatment has little effect upon the cotyledons that are present at the time of irradiation, but damage can be observed as an inhibition of appearance of the true leaves that require cell division to develop (7).

Increasing doses of γ-radiation inhibited the formation of the first rosette leaves. At 10 Krad radiation, both ecotypes produced two rosette leaves, but the size of these leaves was much smaller than that of plants either unexposed or exposed to 5 Krad radiation. At 15 Krad radiation, 64% of the Aa-0 seedlings still produced two small rosette leaves, whereas most of the "WS" seedlings were so severely damaged that no true leaves developed. At 30 Krad radiation neither ecotype produced true leaves (data not shown). By this test, ecotype "WS" plants were approximately 5-6 times more sensitive to γ-radiation than were plants of ecotype Aa-0. This result correlates well with the relative extent of T-DNA integration between these ecotypes.

Genetic basis for differences in susceptibility to crown gall tumorigenesis and other stable transformation phenotypes

To determine whether susceptibility to tumorigenesis is a heritable trait in *Arabidopsis*, we crossed ecotypes Aa-0 and "WS" and determined the pattern of susceptibility to tumorigenesis in subsequent generations. Figure 6A shows that susceptibility of inoculated root segments to tumorigenesis in the F1 generation is a dominant characteristic. These results were verified using a flower bolt inoculation assay. We next determined the pattern of segregation of susceptibility to tumorigenesis in the F2 generation. To do this, we used a semi-quantitative flower bolt inoculation assay in which we measured the weight of bolt segments surrounding the inoculation site 30 days after infection. As shown in Figures 6B and 10, inoculation of flower bolts of ecotype Aa-0 usually resulted in the development of tumors. Most of these tumors were large, weighing greater than 31 mg. Inoculation of flower bolts of ecotype "WS" generally resulted in a wound response, with most tissue segments weighing less than 15-20 mg. Among the F2 segregants, approximately 1/4 of the tumors were small (37/120 [30.8%] of the tumors weighed less than 10 mg); approximately 1/4 of the tumors were large (35/120 [29.2%] of the tumors weighed more than 31 mg), and approximately 1/2 of the tumors (51/120 [42.5%]) were of intermediate size. This approximate 1:2:1 segregation ratio would suggest that susceptibility to tumorigenesis is a semi-dominant trait and apparantly contradicts the root segment inoculation assay results (Figure 6A) that indicate that susceptibility is completely dominant. We ascribe this incomplete dominance to physiological and/or environmental effects upon tumorigenesis using this stem inoculation assay. Alternatively, the results could indicate segregation of multiple loci that together result in susceptibility.

In an attempt to distinguish between these possibilities, we used a different assay to determine the heritability of another stable phenotype, kanamycin-resistance. We inoculated a large number of root bundle segments from individual plants with *A. tumefaciens* GV3101(pBISN1) and selected stable transformants on CIM containing kanamycin. We then calculated the percentage of root bundles from each plant that generated kanamycin-resistant calli. This percentage was 21-

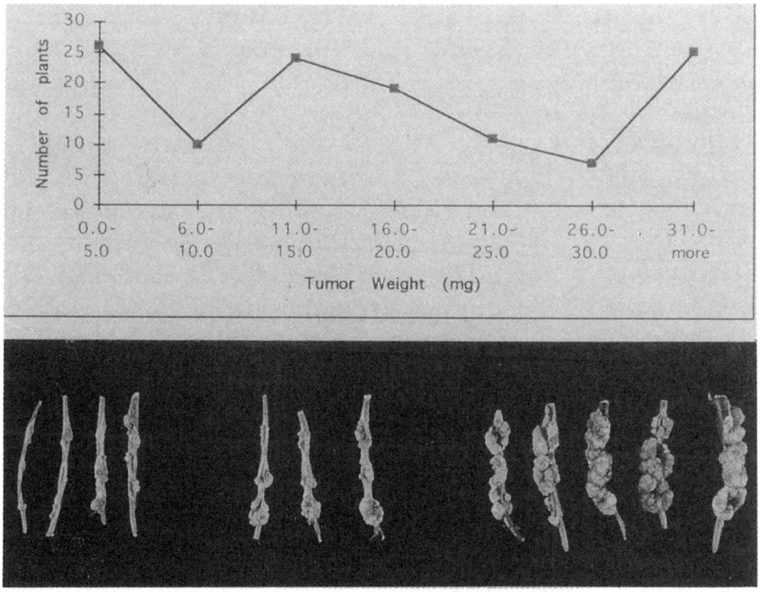

Figure 10. Segregation of phenotype of susceptibility to tumorigenesis in a cross between ecotypes Aa-0 and "WS". The top panel shows the number of small, medium, and large tumors among the F2 progeny. Flower bolts were inoculated *with A. tumefaciens* A208 and the tumors weighed one month later. Bottom, photographs of small, medium, and large tumors.

40% for the recalcitrant ecotype "WS" and greater than 81% for the susceptible ecotype Aa-0 (data not shown). Greater than 81% of the bundles of root segments of the F1 progeny were kanamycin-resistant, indicating that the inheritance of this trait is completely dominant (data not shown). Among the 63 F2 progeny analyzed, 45 plants (71.4%) had root bundles that were highly kanamycin-resistant (>51% of the root bundles from each plant were kanamycin-resistant), whereas 18 plants (28.6%) had root bundles that were mostly kanamycin-susceptible (<40% of the root bundles from each plant were kanamycin-resistant).

Taken together, these data indicate that susceptibility to crown gall tumorigenesis, and the ability to be transformed to kanamycin-resistance, are heritable traits. The data further suggest that a single predominant locus determines the inheritance of these traits. We cannot rule out, however, the possibile influence of other, less predominant loci upon these phenotypes.

REFERENCES

1. **Bergmenn, B.A., and A.M. Stomp.** 1992. Effect of host plant genotype and growth rate on *Agrobacterium tumefaciens*-mediated gall formation in *Pinus radiata.* Phytopathol. **82**:1457-1462.

2. **Christie, P.J., J.E. Ward, S.C. Winans, and E.W. Nester.** 1988. The *Agrobacterium tumefaciens virE2* gene product is a single-stranded-DNA-binding protein that associates with T-DNA. J. Bacteriol. **170**:2659-2667.

3. **Citovsky, V., G. De Vos, and P. Zambryski.** 1988. Single-stranded DNA binding protein encoded by the *virE* locus of *Agrobacterium tumefaciens.* Science **240**:501-504.

4. **Citovsky, V., J. Zupan, D. Warnick, and P. Zambryski.** 1992. Nuclear localization of *Agrobacterium* VirE2 protein in plant cells. Science **256**:1802-1805.

5. **Citovsky, V., D. Warnick, and P. Zambryski.** 1994. Nuclear import of *Agrobacterium* VirD2 and VirE2 proteins in maize and tobacco. Proc. Natl. Acad. Sci. USA **91**:3210-3214.

6. **Das, A.** 1988. *Agrobacterium tumefaciens virE* operon encodes a single-stranded DNA-binding protein. Proc. Natl. Acad. Sci. USA **85**:2909-2913.

7. **Davies, C., D. Howard, G. Tam, and N. Wong.** 1994. Isolation of *Arabidopsis thaliana* mutants hypersensitive to gamma radiation. Mol. Gen. Genet. **243**:660-665.

8. **Filichkin, S.A., and S.B. Gelvin.** 1993. Formation of a putative relaxation intermediate during T-DNA processing directed by the *Agrobacterium tumefaciens* VirD1,D2 endonuclease. Mol. Microbiol. **8**:915-926.

9. **Gheysen, G., R. Villarroel, and M. Van Montagu.** 1991. Illegitimate recombination in plants: A model for T-DNA integration. Genes and Develop. **5**:287-297.

10. **Gietl, C., Z. Koukolikova-Nicola, and B. Hohn.** 1987. Mobilization of T-DNA from *Agrobacterium* to plant cells involves a protein that binds single-stranded DNA. Proc. Natl. Acad. Sci. USA **84**:9006-9010.

11. **Herrera-Estrella, A., Z.-M. Chen, M. Van Montagu, and K. Wang.** 1988 VirD proteins of *Agrobacterium tumefaciens* are required for the formation of a covalent DNA-protein complex at the 5' terminus of T-strand molecules. EMBO J. **7**:4055-4062.

12. **Herrera-Estrella, A., M. Van Montagu, and K. Wang.** 1990. A bacterial peptide acting as a plant nuclear targeting signal: The amino-terminal portion of *Agrobacterium* VirD2 protein directs a β-galactosidase fusion protein into tobacco nuclei. Proc. Natl. Acad. Sci. USA **87**:9534-9537.
13. **Howard, E., V. and Citovsky.** 1990. The emerging structure of the *Agrobacterium* T-DNA transfer complex. BioEssays **12**:103-108.
14. **Howard, E.A., B.A. Winsor, G. De Vos, and P. Zambryski.** 1989. Activation of the T-DNA transfer process in *Agrobacterium* results in the generation of a T-strand-protein complex: Tight association of VirD2 with the 5' ends of T-strands. Proc. Natl. Acad. Sci. USA **86**:4017-4021.
15. **Howard, E.A., J.R. Zupan, V. Citovsky, and P.C. Zambryski.** 1992. The VirD2 protein of *A. tumefaciens* contains a C-terminal bipartite nuclear localization signal: Implications for nuclear uptake of DNA in plant cells. Cell **68**:109-118.
16. **Janssen, B.-J., and R.C. Gardner.** 1989. Localized transient expression of GUS in leaf discs following cocultivation with *Agrobacterium*. Plant Mol. Biol. **14**:61-72.
17. **Jayaswal, R.K., K. Veluthambi, S.B. Gelvin, and J. Slightom.** 1987. Double-stranded cleavage of T-DNA and generation of single-stranded T-DNA molecules in *Escherichia coli* by a *virD*-encoded border-specific endonuclease from *Agrobacterium tumefaciens*. J. Bacteriol. **169**:5035-5045.
18. **Koukolikova-Nicola, Z., D. Raineri, K. Stephens, C. Ramos, B. Tinland, E.W. Nester, and B. Hohn.** 1993. Genetic analyis of the *virD* operon of *Agrobacterium tumefaciens*: A search for functions involved in transport of T-DNA into the plant cell nucleus and in T-DNA integration. J. Bacteriol. **175**:723-731.
19. **Kuldau, G.A., G. De Vos, J. Owen, G. McCaffrey, and P. Zambryski.** 1990. The *virB* operon of *Agrobacterium tumefaciens* pTiC58 encodes 11 open reading frames. Mol. Gen. Genet. **221**:256-266.
20. **Liu, C.-N., X.-Q. Li, and S.B. Gelvin.** 1992. Multiple copies of *virG* enhance the transient transformation of celery, carrot, and rice tissues by *Agrobacterium tumefaciens*. Plant Mol. Biol. **20**:1071-1087.
21. **Matsumoto, S., Y. Ito, T. Sosoi, Y. Takahashi, and Y. Machida.** 1990. Integration of *Agrobacterium* T-DNA

into a tobacco chromosome: Possible involvement of DNA homology between T-DNA and plant DNA. Mol. Gen. Genet. **224**:309-316.

22. **Mayerhofer, R., Z. Koncz-Kalman, C. Nawrath, G. Bakkeren, A. Crameri, K. Angelis, G.R. Redei, J. Schell, B. Hohn, and C. Koncz.** 1991. T-DNA integration: A mode of illegitimate recombination in plants. EMBO J. **10**:697-704.

23. **Narasimhulu, S.B., X.-B. Deng, R. Sarria, and S.B. Gelvin.** 1996. Early transcription of *Agrobacterium* T-DNA genes in tobacco and maize. Plant Cell **8**:873-886.

24. **Ni, M., D. Cui, J. Einstein, S. Narasimhulu, C.E. Vergara, and S.B. Gelvin.** 1995. Strength and tissue specificity of chimeric promoters derived from the octopine and mannopine synthase genes. Plant J. **7**:661-676.

25. **Ohba, T., Y. Yoshioka, C. Machida, and Y. Machida.** 1995. DNA rearrangement associated with the integration of T-DNA into tobacco: An example for multiple duplications of DNA around the integration target. Plant J. **7**:157-164.

26. **Okamoto, S., A. Toyoda-Yamamoto, K. Ito, I. Takebe, and Y. Machida.** 1991. Localization and orientation of the VirD4 protein of *Agrobacterium tumefaciens* in the cell membrane. Mol. Gen. Genet. **228**:24-32.

27. **Rossi, L., B. Hohn, and B. Tinland.** 1993. The VirD2 protein of *Agrobacterium tumefaciens* carries nuclear localization signals important for transfer of T-DNA to plants. Mol. Gen. Genet. **239**:345-353.

28. **Shurvinton, C.E., L. Hodges, and W. Ream.** 1992. A nuclear localization signal and the C-terminal omega sequence in the *Agrobacterium tumefaciens* VirD2 endonuclease are important for tumor formation. Proc. Natl. Acad. Sci. USA **89**:11837-11841.

29. **Stachel, S.E., and E.W. Nester.** 1986. The genetic and transcriptional organization of the *vir* region of the A6 Ti plasmid of *Agrobacterium tumefaciens*. EMBO J. **5**:1445-1454.

30. **Stachel, S.E., B. Timmerman, and P. Zambryski.** 1986. Generation of single-stranded T-DNA molecules during the initial stages of T-DNA transfer from *Agrobacterium tumefaciens* to plant cells. Nature **322**:706-712.

31. **Stachel, S.E., B. Timmerman, and P. Zambryski.** 1987. Activation of *Agrobacterium tumefaciens vir* gene

expression generates multiple single-stranded T-strand molecules from the pTiA6 T-region: Requirement for 5' *virD* gene products. EMBO J. **6**:857-863.

32. **Thompson, D.V., L.S. Melchers, K.B. Idler, R.B. Schilperoort, and P.J.J. Hooykaas.** 1988. Analysis of the complete nucleotide sequence of the *Agrobacterium tumefaciens virB* operon. Nucl. Acids Res. **16**:4621-4636.

33. **Tinland, B., B. Hohn, and H. Puchta.** 1994. *Agrobacterium tumefaciens* transfers single-stranded transferred DNA (T-DNA) into the plant cell nucleus. Proc. Natl. Acad. Sci. USA **91**:8000-8004.

34. **Tinland, B., Z. Koukolikova-Nicola, M.N. Hall, and B. Hohn.** 1992. The T-DNA-linked VirD2 protein contains two distinct functional nuclear localization signals. Proc. Natl. Acad. Sci. USA **89**:7442-7446.

35. **Vancanneyt, G., R. Schmidt, A. O'Connor-Sanchez, L. Willmitzer, and M. Rocha-Sosa.** 1990. Construction of an intron-containing marker gene: Splicing of the intron in transgenic plants and its use in monitoring early events in *Agrobacterium*-mediated plant transformation. Mol. Gen. Genet. **220**:245-250.

36. **Veluthambi, K., W. Ream, and S.B. Gelvin.** 1988. Virulence genes, borders, and overdrive generate single-stranded T-DNA molecules from the A6 Ti plasmid of *Agrobacterium tumefaciens*. J. Bacteriol. **170**:1523-1532.

37. **Ward, E.R., and W.M. Barnes.** 1988. VirD2 protein of *Agrobacterium tumefaciens* very tightly linked to the 5' end of T-strand DNA. Science **242**:927-930.

38. **Ward, J.E., D.E. Akiyoshi, D. Regier, A. Datta, M.P. Gordon, and E.W. Nester.** 1988. Characterization of the *virB* operon from an *Agrobacterium tumefaciens* Ti plasmid. J. Biol. Chem. **263**:5804-5814.

39. **Winans, S.C.** 1992. Two-way chemical signaling in *Agrobacterium*-plant interactions. Microbiol. Rev. **56**:12-31.

40. **Young, C., E.W. and Nester.** 1988. Association of the VirD2 protein with the 5' end of T strands in *Agrobacterium tumefaciens*. J. Bacteriol. **170**:3367-3374.

41. **Yusibov, V.M., T.R. Steck, V. Gupta, and S.B. Gelvin.** 1994. Association of single-stranded transferred DNA from *Agrobacterium tumefaciens* with tobacco cells. Proc. Natl. Acad. Sci. **91**:2994-2998.

CHAPTER 8

VirE1 Protein Mediates Export of the Single-Strand DNA-Binding Protein VirE2 from *Agrobacterium tumefaciens* into Plant Cells

Christopher Sundberg[1], Laura Meek[2], Priscilla Dombek[2], Kate Carroll[3], Anath Das[4], and Walt Ream[1,3]*

Program in Molecular Biology[1], Department of Biochemistry & Biophysics[2], and Department of Agricultural Chemistry[3], Oregon State University, Corvallis, Oregon 97331; Department of Biochemistry and Plant Molecular Genetics Institute, University of Minnesota, St. Paul, Minnesota 55108[4].

*Corresponding author. Telephone: 503-737-1791FAX: 503-737-0497 e-mail: reamw@bcc.orst.edu

ABSTRACT: *Agrobacterium tumefaciens* transfers single-stranded DNAs (T-strands) into plant cells. VirE1 and VirE2, which is a single-stranded DNA binding protein, are important for tumorigenesis. We show that T-strands and VirE2 can enter plant cells independently and that export of VirE2, but not of T-strands, depends on VirE1.

Introduction

Agrobacterium tumefaciens causes crown gall tumors on many dicotyledonous plant species when the bacteria infect wounded tissue (18). The tumor-inducing (Ti) plasmid carries genes essential for tumorigenesis (63,72). The transferred DNA (T-DNA) portion of the Ti plasmid enters plant cells and integrates into nuclear DNA (8,9,74), in which expression of certain T-DNA genes leads to tumorous growth (2,5,32,46,59). Virulence (*vir*) genes necessary for T-DNA transmission (transfer and integration) lie elsewhere on the Ti plasmid (24,55), and wounded plants release phenolic compounds that induce *vir* expression (54).

Border sequences define the T-DNA ends (42,43,48,65). T-DNA transfer begins when the VirD2 endonuclease nicks the right-hand border sequence (66,78) and attaches to the 5' end of the nicked DNA strand (22,27,30,67,79). Strand displacement continues leftward (to the nicked left-hand border sequence) generating linear VirD2-bound T-strands (3,33,56), which the bacteria appear to export into plant cells (58,80). VirE2 single-stranded DNA-binding (SSB) protein (11-13,16,26,47) may also accompany T-strands into plant cells. Both VirE2 and VirD2 contain plant nuclear localization signals (NLSs) and probably play important roles inside infected plant cells (14,28,31,51,62). Export of DNA and proteins from *A. tumefaciens* into plant cells depends upon membrane-associated proteins encoded by the *virB* operon (10,23,34,49,50,60,61,68-71) (for reviews see 29,81) and the *virD4* gene (33,38). The VirB proteins are similar to pertussis toxin liberation (Ptl) proteins of *Bordetella pertussis* that mediate export of pertussis toxin (15,73) and to proteins that facilitate conjugal transfer of IncPα plasmid RP4 (Trb proteins) (35) and IncN plasmid pKM101 (Tra proteins) (44). VirD4 protein has similarity to TraG, another protein required for conjugal transfer of RP4 (35). Thus, *Agrobacterium* proteins essential for tumorigenesis appear to facilitate export of both proteins and DNA into plant cells by using pathways that operate in other bacteria.

VirE2 plays an important role inside plant cells but not inside *Agrobacterium* cells. T-strands (56), which are widely accepted as intermediates of T-DNA transfer (45,75,81), accumulate to wild-type levels in *virE2* mutants (57,64) showing that VirE2 does not stabilize T-strands inside bacterial cells. In addition, *virE2* mutants can transfer T-strands into plant cells (80), albeit with unknown efficiency, proving that VirE2 is not essential for export of T-strands. Transgenic tobacco plants that produce VirE2 protein are susceptible to transformation by a *virE* mutant (14) indicating that VirE2 is necessary inside plant cells.

A. tumefaciens exports VirE2 protein into plant cells via a *vir*-dependent pathway. Otten et al. (39) inoculated two nonpathogenic *A. tumefaciens* strains, a *virE2* mutant and a strain lacking T-DNA, into the same plant wound and observed tumorigenesis. Because these strains were unable to exchange genes by conjugation, they concluded that wild-type *A. tumefaciens* "produce a diffusible factor required for transformation" (39). The *virE2* mutant apparently exported naked T-strands into plant cells and the *virE*$^+$ strain, which lacked T-DNA, exported VirE2 SSB protein. This phenomenon, which is called extracellular complementation, indicates that export of T-DNA and VirE2 protein into plant cells can occur independently. Christie et al. (11) demonstrated that the VirE2 donor must have the VirB proteins, VirD4, and loci (*chvA, chvB*, and *exoC*) necessary for binding of *Agrobacterium* cells to plant cells. The latter observation indicates that the *virE*$^+$ strain probably exports VirE2 directly into plant cells, rather than into the *virE2* mutant.

Tumorigenesis requires both VirE1 and VirE2 (37); however, the function of VirE1 was previously unknown. VirE1 stabilizes VirE2 in *E. coli* suggesting that these proteins interact (37). To determine whether VirE1 also stabilizes VirE2 in *A. tumefaciens* and whether export of VirE2 requires VirE1, we created an in-frame deletion within the *virE1* gene that did not affect *virE2*. The absence of VirE1 protein did not diminish accumulation of VirE2 in *A. tumefaciens* cells. However, the *virE1* mutation prevented export of VirE2, but not T-strands, into plant cells.

Bacterial strains. All *A. tumefaciens* strains were derived from A348, which harbors the octopine-type Ti plasmid pTiA6NC in the C58 chromosomal background (25). *A. tumefaciens* KE1 contains pTiA6ΔE, in which the *virE* region has been removed; the 6.7-kb *virE*-containing region between restriction fragments *Bam*HI 18 and *Eco*RI 11 has been replaced by a kanamycin resistance gene (37). Strain MX358 contains a *virE2*::Tn*3-lacZ* mutation (55); strain WR5001, which was derived from the *chvB*::Tn*5* strain A1038 (24), also contains this same mutation in *virE2*. In pTiA6, border sequences occur at the ends of two adjacent T-DNAs, T_L and T_R, the left- and right-hand T-DNAs. Thus, three border sequences lie to the right of T_L, which carries the genes necessary for tumor growth. Strain WR3095 lacks these three border sequences and therefore is unable to produce T-strands or to transfer T-DNA to plants (42,43).

In-frame deletion in *virE1*. To create a mutant *virE* operon that encoded VirE2 but not VirE1, we removed codons 9 through 62 from *virE1*. The *virE* operon of pTiA6 lies within a 3.2 kb *Xho*I restriction fragment, which we inserted into the *Sal*I site of pUC18 (77) such that *lacZ* and *virE* were transcribed in the same direction in the resulting plasmid, pGR1. We excised the *virE* operon from pGR1 by digestion with *Bam*HI and *Pst*I, purified the 3.2 kb restriction fragment by agarose gel electrophoresis, and inserted it into pUC8 to create pLM1. The *virE* operon has a single *Sph*I site within *virE1* and a single *Stu*I site within *virE2* (76) (Fig. 1). We prepared pLM1 DNA from *E. coli* strain GM119 (*dam-3 dcm-6*) (36) and digested it with *Sph*I and *Stu*I to remove 210 bp from the *virE* operon. The excised region extended from codon 7 of *virE1* through codon 10 of *virE2* (Fig. 1); the remaining linear pLM1 DNA had one *Sph*I-cohesive end and one blunt (*Stu*I-cut) end. Using T4 DNA ligase, we joined an annealed pair of DNA oligonucleotides, flanked by *Sph*I-cohesive and blunt ends, to the *Sph*I and *Stu*I-cut pLM1 DNA. The *Sph*I-compatible end of the annealed oligonucleotides restored codons 7 and 8 of *virE1*. The remainder of the molecule contained the final 3 codons and TGA stop codon of *virE1* which were followed by the 4 bp intergenic region and the first 10 codons of *virE2* (Fig. 1). Thus, substitution of the oligonucleotides shown in Fig. 1 for the *Sph*I-*Stu*I restriction fragment in the *virE* locus created a mutant *virE1* gene (*virE1Δ54*), which contained only 11 codons, without altering *virE2*; this *virE1* mutant plasmid was called pLM2.

A VirE1$^-$ VirE2$^+$ strain. To introduce the *virE1Δ54* mutation into *A. tumefaciens*, we cleaved pLM2 at its single *Bam*HI site (Fig. 1) and inserted it into the *Bam*HI site of the broad-host-range plasmid pRK310, which encodes tetracycline resistance (20), such that the *virE* operon in pLM2 and the *lacZ* gene of pRK310 lie in the same direction. We introduced the resulting plasmid, pKC1, into *A. tumefaciens* KE1, which lacks the *virE* region. KE1(pKC1) is *virE2*$^+$ but mutant for *virE1*.

To create a near-isogenic strain with an intact *virE* operon, we inserted pGR1 (see above) into pRK310 in the same orientation, thereby creating pPD100. KE1(pPD100) is *virE1*$^+$ *virE2*$^+$.

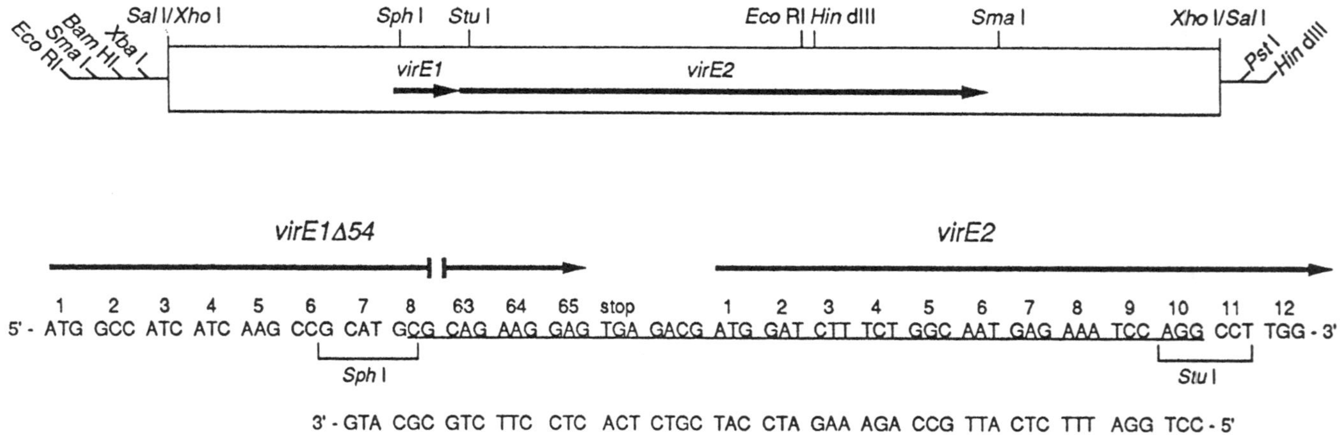

Figure 1. Restriction map of the *virE* operon and sequence of the *virE1Δ54* mutation. Arrows indicate the location of the *virE1* and *virE2* coding sequences. *Sal*I restriction sites within the *virE* region are not shown. The top line of DNA sequence shows the entire *virE1Δ54* gene and the first 12 codons of *virE2*; numbers above the sequence refer to the positions of the codons within the wild-type genes, and the underlined bases comprise one of the oligonucleotides used to create the deletion. The bottom line of DNA sequence shows the complementary oligonucleotide. (Reprinted, by permission, from Sundberg et al. [58a]).

Stability of VirE2 Protein. To determine whether VirE1 protein affects VirE2 accumulation in *A. tumefaciens*, we used standard immunoblot methods (4) to compare VirE2 levels in KE1(pPD100) (*virE1*$^{+}$ *virE2*$^{+}$) with those in KE1(pKC1) (*virE1Δ54 virE2*$^{+}$). Neither strain contained a detectable amount of VirE2 protein in cultures grown without acetosyringone. However, both strains contained a significant amount of VirE2 protein after overnight incubation (24°C) in M9 minimal medium (pH 5.4) containing 100μM acetosyringone (Fig. 2).

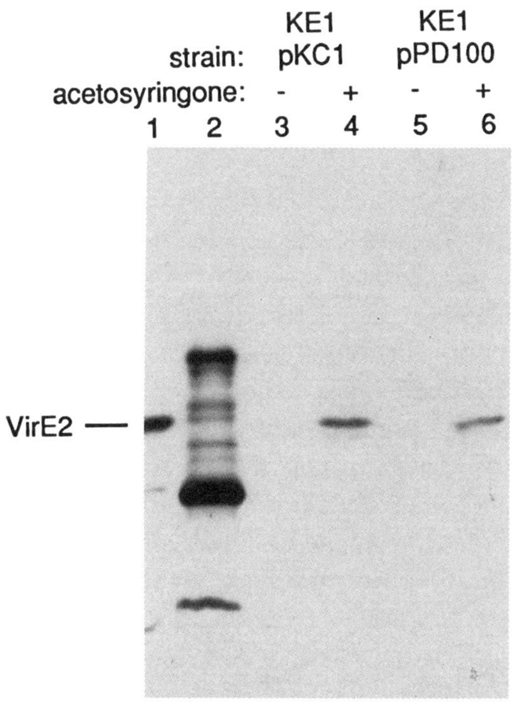

Figure 2. Western blot analysis of VirE2 stability. Lane 1, purified VirE2 protein (20 ng). Lane 2, biotinylated molecular weight standards (Bio-Rad broad-range; 5 μl). Lanes 3 & 4, KE1(pKC1) *cells* (*virE1Δ54 virE2*$^{+}$); lanes 5 & 6, KE1(pPD100) cells (*virE1*$^{+}$ *virE2*$^{+}$). Lanes 3 & 5 contain cells with uninduced *vir* genes, whereas cells in lanes 4 & 6 were treated with acetosyringone to induce *vir* genes. Lanes 3-6 each contained 24 μg of protein. (Reprinted, by permission, from Sundberg et al. [58a]).

A quantitative comparison showed that the amount of VirE2 in KE1(pKC1) was 92% of that in KE1(pPD100). Cells were sonicated, and debris was removed by centrifugation (15,600 x g for 2 min). Protein concentrations in the supernatants were measured by the Bradford dye-binding procedure (7) (Bio-Rad). To detect VirE2, we incubated western blots with polyclonal rabbit antiserum raised against VirE2 (16) and used goat anti-rabbit antiserum conjugated to horseradish peroxidase (Bio-Rad) as a secondary antibody. We detected the bound antibody with the ECL immunoblot chemiluminescence development kit (Amersham). To quantitate light emissions, we used flashed X-ray film and measured band intensity with a Molecular Dynamics densitometer coupled with Image Quant software.

Coinfection tests. Otten et al. discovered that coinoculation of a host plant with two nonpathogenic *A. tumefaciens* strains, one with a mutation in *virE2* and another lacking T-DNA, restored tumorigenesis (39). Thus, coinfection tests provide a means to detect export of VirE2 protein from *A. tumefaciens* into plant cells. We used this method to determine whether export of VirE2 to plant cells requires VirE1. To prepare cultures for inoculation onto plants, we grew *A. tumefaciens* overnight at 28 °C with aeration in YEP broth containing appropriate antibiotics. We washed the bacterial cells, suspended them in phosphate-buffered saline (PBS), and inoculated potato tuber discs (7 mm diameter) as described previously (52); discs inoculated with two strains received 10^7 cells of each. *Kalanchoe daigremontiana* leaves were inoculated with concentrated bacterial suspensions (5×10^8 cells/μl) in PBS; for coinfection of *Kalanchoe*, we mixed strains (1:1) prior to inoculation. We scored tumors after incubation of infected plants at room temperature for 21 days.

VirE1 is essential for export of VirE2. VirE1 plays an important but undetermined role in tumorigenesis (37). Because the *virE1* mutant KE1(pKC1) accumulated normal levels of VirE2 upon induction of the *vir* operons, we asked whether this strain could donate VirE2 to plant cells during coinfection with a nonpathogenic VirE2-deficient strain (MX358), which retains the ability to produce

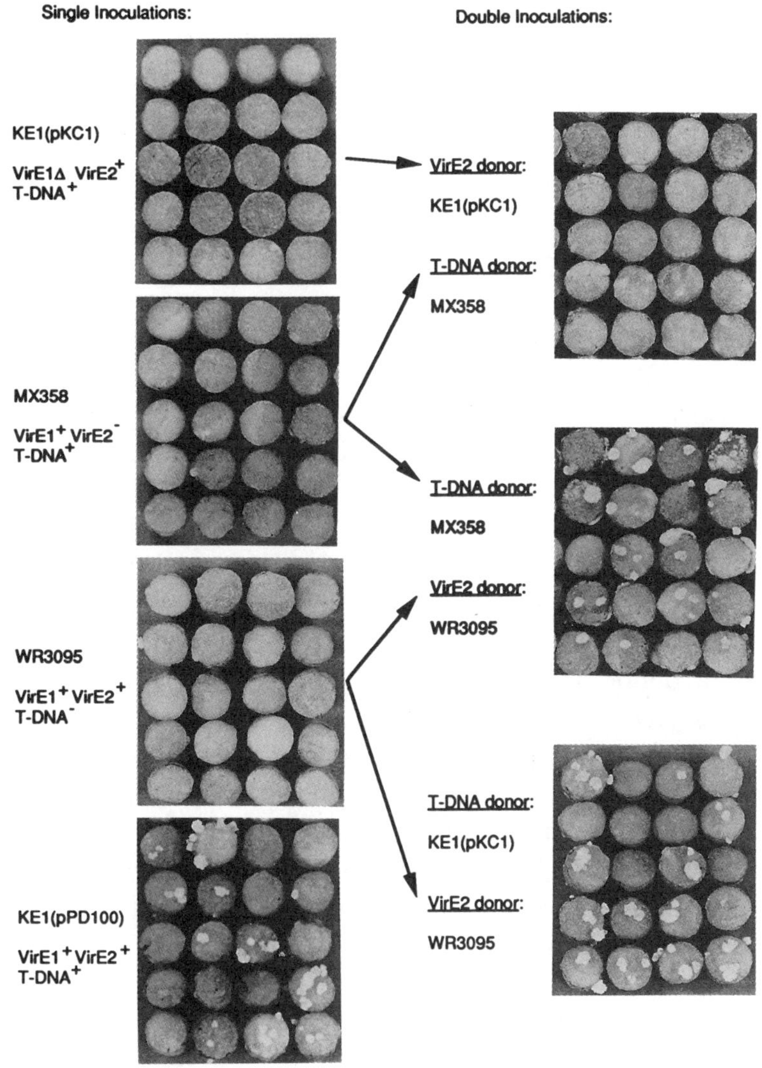

Figure 3. Extracellular complementation. Tumors on potato disks inoculated with one (left column) or two (right column) *A. tumefaciens* strains. Note that KE1(pKC1) is a good T-strand donor (bottom right panel) but an ineffective VirE2 donor (top right panel). (Reprinted, by permission, from Sundberg et al. [58a]).

T-strands (57,64). Figure 3 presents the results of these coinfection tests. As anticipated, the *virE1* mutant did not induce tumors when inoculated onto potato tuber discs, carrot root slices, or leaves of *K. daigremontiana*. The *virE2* mutant also failed to induce tumors on these hosts. Plant wounds coinfected with both *virE1* and *virE2* mutants did not develop tumors. Thus, although a *virE1* mutant can accumulate VirE2, it is unable to restore pathogenicity to a *virE2* mutant by extracellular complementation, indicating that a *virE1* mutant cannot export VirE2 into plant cells.

For comparison, Figure 3 also presents the results of a successful coinfection with the *virE2* mutant MX358 as T-strand donor and WR3095, a *virE*$^+$ strain unable to make T-strands (64), as the VirE2 donor. This coinfection produced tumors at approximately the same frequency as inoculation with a single strain [KE1(pPD100)] that carries both a wild-type *virE* operon and intact T-DNA (Fig. 3).

Coinfection requires plant cell binding. Extracellular complementation requires the VirE2 donor to bind plant cells, and *chvB* is essential for this binding (11). To determine whether the T-strand donor must also bind plant cells for extracellular complementation to occur, we constructed a *chvB*::Tn*5* *virE2*::Tn*3-lacZ* double mutant (WR5001). This strain was unable to donate T-strands in a coinfection experiment with WR3095 as the VirE2 donor; no tumors arose (data not shown). Thus, *chvB*, is needed in both the T-strand and VirE2 donors for extracellular complementation. This suggests that the plant-binding ability of both *Agrobacterium* strains is important.

T-DNA export is independent of VirE1. Mutations in the *virE* operon do not diminish the ability to produce and accumulate T-strands (57,64). Thus, after growth in medium containing acetosyringone, the *virE1* mutant KE1(pKC1) should contain both VirE2 protein (Fig. 2) and T-strands. Several models for T-strand production and transfer postulate that VirE2 protein binds T-strand DNA inside *A. tumefaciens* as soon as the strand is displaced from the nicked Ti plasmid (21,29,45,75,81). Because KE1(pKC1) appears unable to export its VirE2 protein into infected plant cells due to a mutation in *virE1* (Fig. 3), we asked whether VirE2 protein can bind T-strands produced in these cells and thereby prevent or diminish export of T-strands. To test KE1(pKC1) for export of T-strands, we used this strain as T-strand donor in a coinfection experiment with WR3095 as the VirE2 donor. This coinfection resulted in tumorigenesis comparable to that seen on tissue inoculated with both WR3095 and MX358 (Fig. 3). Thus, wild-type levels of VirE2 protein in KE1(pKC1) cells did not interfere with export of T-strands into plant cells.

Discussion

Previous work has shown that VirE1 is important for tumorigenesis and hinted that it may interact directly with VirE2 (37); however, these studies did not define the role of VirE1 in T-DNA transmission. In this study, we show that export of VirE2 into plant cells, as manifested by extracellular complementation, requires VirE1.

In contrast to the *B. pertussis* Ptl proteins, *Agrobacterium* Vir proteins export both DNA and protein. However, transfer of T-DNA did not require VirE1. Although the VirB proteins and VirD4 appear to act in both T-DNA transfer and VirE2 secretion, VirE1 functioned only in secretion of VirE2. Thus, VirE1 appears to convert a DNA export apparatus into one capable of protein export.

Extracellular complementation appears to result from independent transfer of VirE2 protein and T-DNA from separate bacteria directly into plant cells. However, an alternate mechanism has been proposed: T-strands might move from the T-DNA donor into the VirE2 donor via *vir*-promoted conjugation (5a), form a complex with VirE2, and then enter plant cells as a complex (Karla Fullner, personal communication). Such a mechanism would require only the VirE2 donor to bind plant cells; however, we have shown that a *chvB* mutation in the T-strand donor prevents extracellular complementation. Also, Fullner et al. showed that a *chvB* mutation does not diminish *vir*-promoted conjugation (Chapter 2, this volume). Because the *chvB* mutation abolishes binding to plant cells without disrupting the potential for conjugation between bacterial cells, we infer that proper attachment of the T-strand donor to plant cells is important for extracellular complementation.

The ability of the *virE1Δ54* strain to transfer T-DNA into plant cells calls into question current models for T-DNA transmission which propose that VirE2 protein binds cooperatively to T-strands inside bacterial cells (11-14,16,26,29,47), especially since experimental support for this hypothesis is limited. We showed here that *virE1* mutants accumulated VirE2 upon induction of the *vir* operons; however, these cells were unable to export this protein into plant cells in extracellular complementation experiments. Citovsky et al. (12) estimated that *vir*-induced *Agrobacterium* cells contain sufficient VirE2, the most abundant Vir protein, to coat all of the T-strands produced. KE1(pKC1) probably contains several times the normal amount of VirE2, because it carries the *virE2* gene on a multicopy plasmid. If VirE2 coats T-strands inside *Agrobacterium* cells as most models predict, we expect that little or no T-strand DNA will remain unbound by VirE2, especially in a strain that produces excess VirE2. Because the *virE1* mutant cannot export VirE2, we infer that it probably cannot export T-strands coated with VirE2. However, we found that VirE2 protein within the *virE1* mutant did not interfere with its ability to transfer T-DNA during extracellular complementation experiments. Thus, these bacteria appear to export uncoated T-strands, even though the cells contain normal amounts of VirE2.

Production and export of T-strands may occur concomitantly at the bacterial membrane, as does nicking and transfer of F plasmid DNA during conjugation (1,17,19,40,41,53). F plasmid encodes an inner membrane protein, TraD, which plays a direct role in DNA translocation (40,41) and also appears to anchor TraI, an *oriT*-specific endonuclease, to the inner membrane (17). VirD4, an inner membrane protein with similarities to F-encoded TraD and RP4 TraG (35,75), may anchor the VirD2 endonuclease to a membrane-associated pore composed of Vir proteins. If so, *Agrobacterium* cells may export naked T-strands into plant cells as they are displaced from the Ti plasmid, before VirE2 has an opportunity to bind. While the present paper was under review, Binns et al. (6) proposed a similar model based on the ability of plasmid RSF1010 to block VirE2 export but not T-strand transfer.

T-strands may accumulate inside acetosyringone-treated bacteria because recipient plant cells are absent. Although VirE2::T-strand complexes were immunoprecipitated from *Agrobacterium* extracts (11), the complex had ample opportunity to form after the cells had been lysed. The conditions used to induce T-strand production do not mimic natural infections, and the methods used to isolate T-strands disrupt the integrity of bacterial cells and make their natural state in vivo difficult to ascertain.

Successful T-DNA transmission does not require that T-strands and VirE2 protein enter plant cells as a complex. Extracellular complementation leads to efficient induction of tumors even though VirE2 and T-strands enter host cells from separate bacteria (11,39). Transgenic tobacco plants that synthesize VirE2 protein restore pathogenicity to a *virE* mutant (14). These observations show that VirE2 and T-strands can form a complex in plant cells and that these molecules need not interact in *Agrobacterium* cells for stable T-DNA transmission to occur. We suggest that separate transfer of VirE2 and T-strands is probably the norm, even during infections with wild-type bacteria.

Acknowledgments

We thank Stanton B. Gelvin for valuable comments on this work, which was supported by USDA grant 93-37301-8728 to W. Ream and US Public Health Service grant GM-37555 to A. Das. C. Sundberg was supported by a USDA National Needs Fellowship.

References

1. **Achtman, M., P.A. Manning, C. Edelbluth, and P. Herrlich**. 1979. Export without proteolytic processing of inner and outer membrane proteins encoded by F sex factor *tra* cistron in *Escherichia coli* minicells. Proc. Natl. Acad. Sci. USA **76**:4837-4841.

2. **Akiyoshi, D.E., H. Klee, R.M. Amasino, E.W. Nester, and M.P. Gordon**. 1984. T-DNA of *Agrobacterium tumefaciens* encodes an

enzyme of cytokinin biosynthesis. Proc. Natl. Acad. Sci. USA **81**:5994-5998.

3. **Albright, L.M., M.F. Yanofsky, B. Leroux, D. Ma, and E.W. Nester**. 1987. Processing of the T-DNA of *Agrobacterium tumefaciens* generates border nicks and linear, single-stranded T-DNA. J. Bacteriol. **169**:1046-1055.

4. **Ausubel, F.M., R. Brent, R.E. Kingston, D.D. Moore, J.G. Seidman, J.A. Smith, and K. Struhl**. 1987. Current protocols in molecular biology, Wiley, New York.

5. **Barry, G.F., S.G. Rogers, R.T. Fraley, and L. Brand**. 1984. Identification of a cloned cytokinin biosynthetic gene. Proc. Natl. Acad. Sci.USA **81**:4776-4780.

5a. **Beijersbergen, A., Dulk-Ras, A., Schilperoort, R.A., and P. Hooykaas**. 1992. Conjugative transfer by the virulence system of *Agrobacterium tumefaciens*. Science **256**:1324-1326.

6. **Binns, A.N., C.E. Beaupre, and E.M. Dale**. 1995. Inhibition of VirB-mediated transfer of diverse substrates from *Agrobacterium tumefaciens* by the IncQ plasmid RSF1010. J. Bacteriol. **177**:4890-4899.

7. **Bradford, M**. 1976. A rapid and sensitive method for the quantitation of microgram quantities of proteins utilizing the principle of protein-dye binding. Anal. Biochem. **72**:248-254.

8. **Chilton, M.D., M.H. Drummond, D.J. Merlo, D. Sciaky, A.L. Montoya, M.P. Gordon, and E.W. Nester**. 1977. Stable incorporation of plasmid DNA into higher plant cells: the molecular basis of crown gall tumorigenesis. Cell **11**:263-271.

9. **Chilton, M.D., R.K. Saiki, N. Yadav, M.P. Gordon, and F. Quetier**. 1980. T-DNA from *Agrobacterium* Ti plasmid is in the nuclear DNA fraction of crown gall tumor cells. Proc. Natl. Acad. Sci. USA 77:4060-4064.

10. **Christie, P.J., J.E. Ward, M.P. Gordon, and E.W. Nester**. 1989. A gene required for transfer of T-DNA to plants encodes an ATPase with autophosphorylating activity. Proc.Natl.Acad.Sci.USA **86**:9677-9681.

11. **Christie, P.J., J.E. Ward, S.C. Winans, and E.W. Nester**. 1988. The *Agrobacterium tumefaciens virE2* gene product is a single-stranded-DNA-binding protein that associates with T-DNA. J.

Bacteriol. **170**:2584-2591.

12. **Citovsky, V., G. De Vos, and P. Zambryski**. 1988. Single-stranded DNA binding protein encoded by the *virE* locus of *Agrobacterium tumefaciens*. Science **240**:501-504.

13. **Citovsky, V., M.L. Wong, and P. Zambryski**. 1989. Cooperative interaction of *Agrobacterium* VirE2 protein with single-stranded DNA: implications for the T-DNA transfer process. Proc. Natl. Acad. Sci. USA **86**:1193-1197.

14. **Citovsky, V., J. Zupan, D. Warnick, and P. Zambryski**. 1992. Nuclear localization of *Agrobacterium* VirE2 protein in plant cells. Science **256**:1802-1805.

15. **Covacci, A., and R. Rappuoli**. 1993. Pertussis toxin export requires accessory genes located downstream from the pertussis toxin operon. Mol. Microbiol. **8**:429-434.

16. **Das, A**. 1988. *Agrobacterium tumefaciens virE* operon encodes a single-stranded DNA-binding protein. Proc. Natl. Acad. Sci. USA **85**:2909-2913.

17. **Dash, P.K., B.A. Traxler, M.M. Panicker, D.D. Hackney, and E.G. Minkley**. 1992. Biochemical characterization of *Escherichia coli* DNA helicase I. Mol. Microbiol. **6**:1163-1172.

18. **DeCleene, M., and J. DeLey**. 1976. The host range of crown gall. Bot. Rev. **42**:389-466.

19. **Di Laurenzio, L., L.S. Frost, and W. Paranchych**. 1992. The TraM protein of the conjugative plasmid F binds to the origin of transfer of the F and ColE1 plasmids. Mol. Microbiol. **6**:2951-2959.

20. **Ditta, G., T. Schmidhauser, E. Yakobson, P. Lu, X. Liang, D.R. Finlay, D. Guiney, and D.R. Helinski**. 1985. Plasmids related to the broad-host-range vector, pRK290, useful for gene cloning and for monitoring gene expression. Plasmid **13**:149-153.

21. **Dreiseikelmann, B**. 1994. Translocation of DNA across bacterial membranes. Micro. Revs. **58**:293-316.

22. **Durrenberger, F., A. Crameri, B. Hohn, and Z. Koukolikova-Nicola**. 1989. Covalently bound VirD2 protein of *Agrobacterium tumefaciens* protects the T-DNA from exonucleolytic degradation. Proc.Natl.Acad.Sci.USA **86**:9154-9158.

23. **Engstrom, P., P. Zambryski, M. Van Montagu, and S.E. Stachel**. 1987. Characterization of *Agrobacterium tumefaciens*

virulence proteins induced by the plant factor acetosyringone. J. Mol. Biol. **197**:635-645.

24. **Garfinkel, D.J., and E.W. Nester**. 1980. *Agrobacterium tumefaciens* mutants affected in crown gall tumorigenesis and octopine catabolism. J. Bacteriol. **144**:732-743.

25. **Garfinkel, D.J., R.B. Simpson, L.W. Ream, F.F. White, M.P. Gordon, and E.W. Nester**. 1981. Genetic analysis of crown gall: fine structure map of the T-DNA by site-directed mutagenesis. Cell **27**:143-153.

26. **Gietl, C., Z. Koukolikova-Nicola, and B. Hohn**. 1987. Mobilization of T-DNA from *Agrobacterium* to plant cells involves a protein that binds single-stranded DNA. Proc.Natl.Acad.Sci.USA **84**:9006-9010.

27. **Herrera-Estrella, A., Z. Chen, M. Van Montagu, and K. Wang**. 1988. VirD proteins of *Agrobacterium tumefaciens* are required for the formation of a covalent DNA-protein complex at the 5' terminus of T-strand molecules. EMBO J. 7:4055-4062.

28. **Herrera-Estrella, A., M. Van Montagu, and K. Wang**. 1990. A bacterial peptide acting as a plant nuclear targeting signal: the amino-terminal portion of *Agrobacterium* Variate protein directs a beta-galactosidase fusion protein into tobacco nuclei. Proc. Natl. Acad. Sci. USA **87**:9534-9537.

29. **Hooykaas, P.J.J., and G.M. Beijersbergen**. 1994. The virulence system of *Agrobacterium tumefaciens*. Ann. Rev. Phytopathol. **32**:157-179.

30. **Howard, E.A., B.A. Winsor, G. De Vos, and P. Zambryski**. 1989. Activation of the T-DNA transfer process in *Agrobacterium* results in the generation of a T-strand-protein complex: tight association of Variate with the 5' ends of T-strands. Proc. Natl. Acad. Sci. USA **86**:4017-4021.

31. **Howard, E.A., J.R. Zupan, V. Citovsky, and P.C. Zambryski**. 1992. The Variate protein of *Agrobacterium tumefaciens* contains a C-terminal bipartite nuclear localization signal: implications for nuclear uptake of DNA in plant cells. Cell **68**:109-118.

32. **Inze, D., A. Follin, M. Van Lijsebettens, C. Simoens, C. Genetello, M. Van Montagu, and J. Schell**. 1984. Genetic analysis of the individual T-DNA genes of *Agrobacterium tumefaciens*; further

evidence that two genes are involved in indole-3-acetic acid synthesis. Mol. Gen. Genet. **194**:265-274.

33. **Jayaswal, R.K., K. Veluthambi, S.B. Gelvin, and J.L. Slightom**. 1987. Double-stranded T-DNA cleavage and the generation of single-stranded T-DNA molecules in *E. coli* by a *virD* encoded border specific endonuclease from *Agrobacterium tumefaciens*. J. Bacteriol. **169**:5035-5045.

34. **Kuldau, G.A., G. De Vos, J. Owen, G. McCaffrey, and P. Zambryski**. 1990. The *virB* operon of *Agrobacterium tumefaciens* pTiC58 encodes 11 open reading frames. Mol. Gen. Genet. **221**:256-266.

35. **Lessl, M., D. Balzer, W. Pansegrau, and E. Lanka**. 1992. Sequence similarities between the RP4 Tra2 and the Ti VirB region strongly support the conjugation model for T-DNA transfer. J. Biol. Chem. **267**:20471-20480.

36. **Marinus, M.G**. 1984. Methylation of procaryotic DNA, p.81-109. In A. Razin, M. Cedar, and A.D. Riggs (eds.), DNA methylation. Springer, New York.

37. **McBride, K.E., and V.C. Knauf**. 1988. Genetic analysis of the *virE* operon of the *Agrobacterium* Ti plasmid pTiA6. J. Bacteriol. **170**:1430-1437.

38. **Okamoto, S., A. Toyoda-Yamamoto, K. Ito, I. Takebe, and Y. Machida**. 1991. Localization and orientation of the VirD4 protein of *Agrobacterium tumefaciens* in the cell membrane. Mol. Gen. Genet. **228**:24-32.

39. **Otten, L., H. DeGreve, J. Leemans, R. Hain, P.J.J. Hooykaas, and J. Schell**. 1984. Restoration of virulence of *vir* region mutants of *Agrobacterium tumefaciens* strain B653 by coinfection with normal and mutant *Agrobacterium* strains. Mol. Gen. Genet. **175**:159-163.

40. **Panicker, M.M., and E.G. Minkley**. 1985. DNA transfer occurs during a cell surface contact stage of F sex factor-mediated bacterial conjugation. J. Bacteriol. **162**:584-590.

41. **Panicker, M.M., and E.G. Minkley**. 1992. Purification and properties of the F sex factor TraD protein, an inner membrane conjugal transfer protein. J. Biol. Chem. **267**:12761-12766.

42. **Peralta, E.G., R. Hellmiss, and W. Ream**. 1986. *Overdrive*, a

T-DNA transmission enhancer on the *A. tumefaciens* tumour-inducing plasmid. EMBO J. **5**:1137-1142.

43. **Peralta, E.G., and L.W. Ream**. 1985. T-DNA border sequences required for crown gall tumorigenesis. Proc.Natl.Acad.Sci.USA **82**:5112-5116.

44. **Pohlman, R.F., H.D. Genetti, and S.C. Winans**. 1994. Common ancestry between IncN conjugal transfer genes and macromolecular export systems of plant and animal pathogens. Mol. Microbiol. **14**:655-668.

45. **Ream, W**. 1989. *Agrobacterium tumefaciens* and interkingdom genetic exchange. Ann. Rev. Phytopathol. **27**:583-618.

46. **Schroeder, G., S. Waffenschmidt, E.W. Weiler, and J. Schroeder**. 1984. The T-region of Ti plasmids codes for an enzyme synthesizing indole-3-acetic acid. Eur. J. Biochem. **138**:387-391.

47. **Sen, P., G.J. Pazour, D. Anderson, and A. Das. 1989**. Cooperative binding of the VirE2 protein to single-stranded DNA. J. Bacteriol. **171**:2573-2580.

48. **Shaw, C.H., M. Watson, and G. Carter**. 1984. The right hand copy of the nopaline Ti plasmid 25 bp repeat is required for tumour formation. Nucleic Acids Res. **12**:6031-6041.

49. **Shirasu, K., and C.I. Kado**. 1993. Membrane location of the Ti plasmid VirB proteins involved in the biosynthesis of a pilin-like conjugative structure on *Agrobacterium tumefaciens*. FEMS Microbiol. Lett. **111**:287-294.

50. **Shirasu, K., P. Morel, and C.I. Kado**. 1990. Characterization of the *virB* operon of an *Agrobacterium tumefaciens* Ti plasmid: nucleotide sequence and protein analysis. Mol. Microbiol. **4**:1153-1163.

51. **Shurvinton, C.E., L. Hodges, and W. Ream**. 1992. A nuclear localization signal and the C-terminal omega sequence in the *Agrobacterium tumefaciens* VirD2 endonuclease are important for tumor formation. Proc.Natl.Acad.Sci.USA **89**:11837-11841.

52. **Shurvinton, C.E., and W. Ream**. 1991. Stimulation of *Agrobacterium tumefaciens* T-DNA transfer by overdrive depends on a flanking sequence but not on helical position with respect to the border repeat. J. Bacteriol. **173**:5558-5563.

53. **Silverman, P.M**. 1987. The structural basis of procaryotic DNA

transfer, p.277-309. In M. Inouye (ed.), Bacterial outer membranes as model systems. Wiley, New York.

54. **Stachel, S.E., E. Messens, M. Van Montagu, and P. Zambryski**. 1985. Identification of the signal molecules produced by wounded plant cells that activate T-DNA transfer in *Agrobacterium tumefaciens*. Nature **318**:624-629.

55. **Stachel, S.E., and E.W. Nester**. 1986. The genetic and transcriptional organization of the *vir* region of the A6 Ti plasmid of *Agrobacterium tumefaciens*. EMBO J. **5**:1445-1454.

56. **Stachel, S.E., B. Timmerman, and P. Zambryski**. 1986. Generation of single-stranded T-DNA molecules during the initial stages of T-DNA transfer from *Agrobacterium tumefaciens* to plant cells. Nature **322**:706-712.

57. **Stachel, S.E., B. Timmerman, and P. Zambryski**. 1987. Activation of *Agrobacterium tumefaciens vir* gene expression generates multiple single-stranded T-strand molecules from the pTiA6 T-region: requirement of 5' *virD* gene products. EMBO J. **6**:857-863.

58. **Stachel, S.E., and P. Zambryski**. 1986. *Agrobacterium tumefaciens* and the susceptible plant cell: A novel adaptation of extracellular recognition and DNA conjugation. Cell **47**:155-157.

58a. **Sundberg, C., Meek, L., Carroll, K., Das, A., and W. Ream**. 1996. VirE1 protein mediates export of the single-stranded DNA-binding protein VirE2 from *Agrobacterium tumefaciens* into plant cells. J. Bacteriol. **178**:1207-1212.

59. **Thomashow, L.S., S. Reeves, and M.F. Thomashow**. 1985. Crown gall oncogenesis: evidence that a T-DNA gene from the *Agrobacterium* Ti plasmid pTiA6 encodes an enzyme that catalyzes synthesis of indoleacetic acid. Proc.Natl.Acad.Sci.USA **81**:5071-5075.

60. **Thompson, D.V., L.S. Melchers, K.B. Idler, R.A. Schilperoort, and P.J.J. Hooykaas**. 1988. Analysis of the complete nucleotide sequence of the *Agrobacterium tumefaciens virB* operon. Nucleic Acids Res. **16**:4621-4636.

61. **Thorstenson, Y.R., G.A. Kuldau, and P.C. Zambryski**. 1993. Subcellular localization of seven VirB proteins of *Agrobacterium tumefaciens*: implications for the formation of a T-DNA transport structure. J. Bacteriol. **175**:5233-5241.

62. **Tinland, B., Z. Koukolikova-Nicola, M.N. Hall, and B. Hohn** 1992. The T-DNA-linked VirD2 protein contains two distinct functional nuclear localization signals. Proc.Natl.Acad.Sci.USA **89**:7442-7446.

63. **Van Larebeke, N., G. Engler, M. Holsters, S. Van den Elsacker, I. Zaenen, R.A. Schilperoort, and J. Schell**. 1974. Large plasmid in *Agrobacterium tumefaciens* essential for crown gall inducing activity. Nature **252**:169-170.

64. **Veluthambi, K., W. Ream, and S.B. Gelvin**. 1988. Virulence genes, borders, and overdrive generate single-stranded T-DNA molecules from the A6 Ti plasmid of *Agrobacterium tumefaciens*. J. Bacteriol. **170**:1523-1532.

65. **Wang, K., L. Herrera-Estrella, M. Van Montagu, and P. Zambryski**. 1984. Right 25 bp terminus sequence of the nopaline T-DNA is essential for and determines direction of DNA transfer from *Agrobacterium* to the plant genome. Cell **38**:455-462.

66. **Wang, K., S.E. Stachel, B. Timmerman, M. Van Montagu, and P. Zambryski**. 1987. Site-specific nick in the T-DNA border sequence as a result of *Agrobacterium vir* gene expression. Science **235**:587-591.

67. **Ward, E.R., and W.M. Barnes**. 1988. VirD2 protein of *Agrobacterium tumefaciens* very tightly linked to the 5' end of T-strand DNA. Science **242**:927-930.

68. **Ward, J.E., D.E. Akiyoshi, D. Regier, A. Datta, M.P. Gordon, and E.W. Nester**. 1988. Characterization of the *virB* operon from an *Agrobacterium tumefaciens* Ti plasmid. J. Biol. Chem. **263**:5804-5814.

69. **Ward, J.E., E.M. Dale, and A.N. Binns**. 1991. Activity of the *Agrobacterium* T-DNA transfer machinery is affected by *virB* gene products. Proc.Natl.Acad.Sci.USA **88**:9350-9354.

70. **Ward, J.E., E.M. Dale, P.J. Christie, E.W. Nester, and A.N. Binns**. 1990. Complementation analysis of *Agrobacterium tumefaciens* Ti plasmid *virB* genes by use of a *vir* promoter expression vector: *virB9*, *virB10*, and *virB11* are essential virulence genes. J. Bacteriol. **172**:5187-5199.

71. **Ward, J.E., E.M. Dale, E.W. Nester, and A.N. Binns**. 1990. Identification of a VirB10 protein aggregate in the inner membrane of

Agrobacterium tumefaciens. J. Bacteriol. **172**:5200-5210.
72. **Watson, B., T.C. Currier, M.P. Gordon, M.D. Chilton, and E.W. Nester**. 1975. Plasmid required for virulence of *Agrobacterium tumefaciens*. J. Bacteriol. **123**:255-264.
73. **Weiss, A.A., F.D. Johnson, and D.L. Burns**. 1993. Molecular characterization of an operon required for pertussis toxin secretion. Proc.Natl.Acad.Sci.USA **90**:2970-2974.
74. **Willmitzer, L., M. Debeuckeleer, M. Lemmers, M. Van Montagu, and J. Schell**. 1980. DNA from Ti plasmid present in nucleus and absent from plastids of crown gall plant cells. Nature **287**:359-361.
75. **Winans, S.C**. 1992. Two-way chemical signaling in *Agrobacterium*-plant interactions. Microbiol. Rev. **56**:12-31.
76. **Winans, S.C., P. Allenza, S.E. Stachel, K.E. McBride, and E.W. Nester**. 1987. Characterization of the *virE* operon of the *Agrobacterium* Ti plasmid pTiA6. Nucleic Acids Res. **15**:825-838.
77. **Yanisch-Perron, C., J. Vieira, and J. Messing**. 1985. Improved M13 phage cloning vectors and host strains: nucleotide sequence of the M13 mp18 and pUC19 vectors. Gene **33**:103-119.
78. **Yanofsky, M.F., S.G. Porter, C. Young, L.M. Albright, M.P. Gordon, and E.W. Nester**. 1986. The *virD* operon of *Agrobacterium tumefaciens* encodes a site-specific endonuclease. Cell **47**:471-477.
79. **Young, C., and E.W. Nester**. 1988. Association of the Variate protein with the 5' end of T strands in *Agrobacterium tumefaciens*. J. Bacteriol. **170**:3367-3374.
80. **Yusibov, V.M., T.R. Steck, V. Gupta, and S.B. Gelvin**. 1994. Association of single-stranded transferred DNA from *Agrobacterium tumefaciens* with tobacco cells. Proc.Natl.Acad.Sci.USA **91**:2994-2998.
81. **Zambryski, P.C**. 1992. Chronicles from the *Agrobacterium*-plant cell DNA transfer story. Annu. Rev. Plant Physiol. Plant Mol. Biol. **43**:465-490.

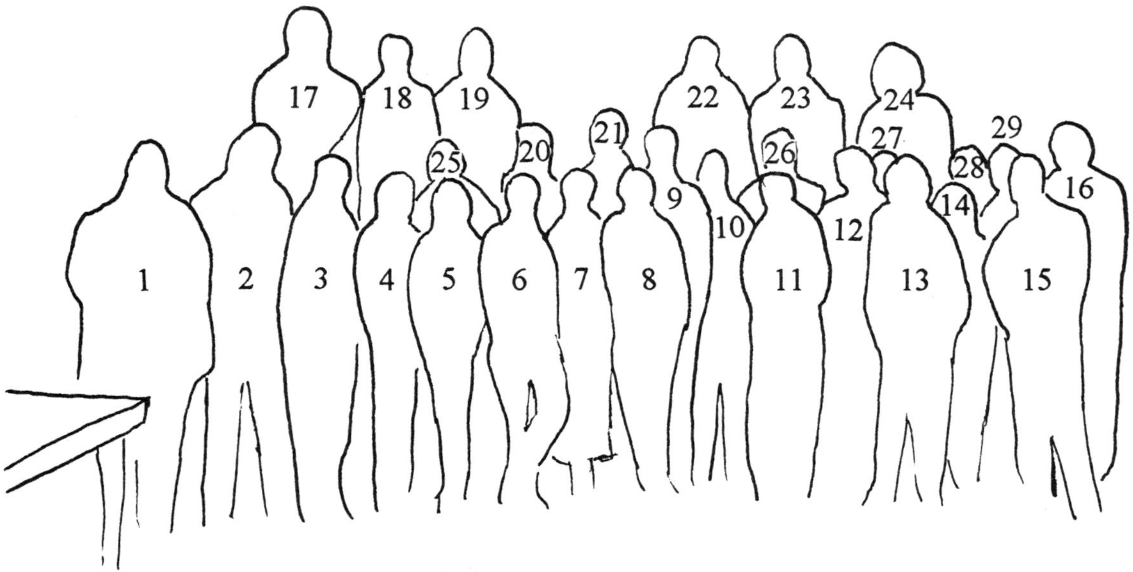
17
18
19
22
23
24
29
27
21
20
26
25
28
16
9
14
10
12
1
2
3
4
5
6
7
8
11
13
15

Key to group photograph (courtesy of Samson Au)

1	Beverly Green
2	Milton Gordon
3	Lorraine Anderson
4	Xiaoyou Liang
5	Sharon Lafferty Doty
6	Karla Fullner
7	Lois Banta
8	Patricia Zambryski
9	Patrice Dion
10	Stan Gelvin
11	Barbara Hohn
12	John Kemner
13	Mary-Dell Chilton
14	Ann Powell
15	Carl Douglas
16	Eugene Nester
17	Samson Au
18	Shouguang Jin
19	Shen Pan
20	Steve Winans
21	Scott Chilton
22	Yong-Woog Lee
23	Wanyin Deng
24	Elizabeth Hood
25	Peter Christie
26	Christian Baron
27	Jeanne Sandahl
28	Walt Ream
29	Anath Das